THE BASICS OF ENERGY AND REACTIONS

CORE CONCEPTS

THE BASICS OF ENERGY AND REACTIONS

CHRIS COOPER

ROSEN PUBLISHING
New York

This edition published in 2014 by:

The Rosen Publishing Group, Inc.
29 East 21st Street
New York, NY 10010

Library of Congress Cataloging-in-Publication Data

Cooper, Christopher (Christopher E.), author.
The basics of energy and reactions/Chris Cooper.—First edition.
pages cm.—(Core concepts)
Audience: Grades 7 to 12.
Includes bibliographical references and index.
ISBN 978-1-4777-2710-2 (library binding)
1. Chemical reactions—Juvenile literature. 2. Matter—Properties—Juvenile literature. I. Title.
QD501.C784 2014
541.39—dc23

2013024900

Manufactured in the United States of America

CPSIA Compliance Information: Batch #W14YA: For further information, contact Rosen Publishing, New York, New York, at 1-800-237-9932.

CONTENTS

CHAPTER ONE

UNDERSTANDING ENERGY IN CHEMICAL REACTIONS

When one substance changes into another, chemical reactions are happening. Some reactions take in energy, but others can give out energy in large amounts.

In the world around us, we can often see changes happening in familiar objects. Shiny metal objects may become dull and tarnished, and food may "go bad." Many of these changes involve chemical reactions.

Burning is a type of chemical reaction. In a forest fire, the burning trees gradually change into ash, smoke, and flames (which are hot, glowing gases). At the same time heat is produced, making the ash, smoke, and gases hot.

Substances change in chemical reactions because atoms, the tiny particles with which matter is made, separate and rejoin in new ways. Wood, for example, is made up of a mixture of chemical compounds

Forest fires are chemical reactions that often occur in hot, dry weather. Once the fire starts, it can take weeks to die down.

TAKING A CLOSER LOOK AT CHEMICAL CHANGES

Chemical reactions are just one type of chemical change. When sugar dissolves in a cup of hot coffee, this, too, is a chemical change. The sugar seems to disappear, but really, the groups of atoms that make up the sugar simply spread out through the liquid. The sugar does not change into anything else: it is just mixed very thoroughly with the liquid. So although it is a change, it is not usually called a reaction because the sugar does not change into a different compound.

containing atoms of carbon and hydrogen, among others. During burning, these atoms separate and join with oxygen atoms from the air to form new compounds: the gases carbon dioxide and carbon monoxide, and water. The atoms themselves have changed very little in the process, but the ways in which they are grouped have changed.

UNDERSTANDING CHEMICAL CHANGE

Chemists study why and how chemical compounds react with each other. They also investigate the effect of changes in temperature, pressure, and other conditions. They try to explain why some

When iron is exposed to water and air for some time, it rusts. A brown iron compound is formed in this reaction.

reactions happen faster than others, and why some need flames, sparks, or other types of extra help to get them started.

For example, the compound methane, which is made up of carbon and hydrogen atoms, can be burned in a kitchen oven to provide heat. During burning, the carbon and hydrogen atoms join with oxygen atoms from the air, forming carbon dioxide and water. Once the burning has been started with a match or a switch on the oven, it will continue as long as the supply of methane and air lasts.

Iron also combines with oxygen in the air. An old iron nail, left out in the weather, becomes rusty. The rust consists of iron combined with oxygen. This reaction is similar to the burning of wood or methane, but happens much more slowly. It also needs no spark or flame to start it, unlike burning methane.

How fast a reaction happens depends on temperature. Almost all chemical reactions go faster when the temperature is higher. The higher the temperature at which we cook our food, the faster it cooks. Separating dirt from clothes is also a chemical process, and it, too, goes better in hot water. Chemists often use heat in the laboratory to make reactions take place more rapidly.

THE PROCESS OF ENERGY AND CHEMICAL CHANGE

The scientific idea of energy can explain a great deal about chemical reactions. In general, the energy that an object has is a

This historical illustration shows the French inventor Denis Papin (1647–1712) with his original pressure cooker. Papin found that increasing the pressure inside a cooking pot makes the contents reach a higher temperature and thus cook more quickly.

A bullet at rest is harmless, but once it is moving at high speed, its movement energy makes it smash through anything in its path, causing damage.

measure of its ability to make things happen. For example, a fast-moving ball can break a pane of glass, make a hole in the ground, or knock over a group of pins in a bowling alley. The ball has energy to do these things because it is moving, and it gives up some of its energy of movement when it makes something happen. This type of energy is called kinetic energy, from the Greek word *kinesis* meaning "movement."

Another example is a lightbulb giving out energy in the form of light and heat. This energy comes from the electric current that flows through the lightbulb. The light makes light-sensitive cells in our eyes react, and this starts the process of seeing. The lightbulb also gives out invisible heat radiation, which warms objects nearby.

In a chemical reaction, the energy in the substances is part of what makes the reaction happen. It also influences how the reaction happens.

MICROSCOPIC MATTER

In explaining chemical change, chemists use their knowledge of how tiny atoms behave. All matter is made of atoms, and these often form groups bonded together, called molecules. Most—but not all—molecules are minute.

In the air around us, molecules of nitrogen and oxygen (the two main gases in the air) are pairs of atoms—although some oxygen exists as ozone, with three atoms. Carbon dioxide consists of one carbon atom linked to two oxygen atoms, while some other gases, such as argon and krypton, consist of single atoms only.

Groups of atoms may break up during chemical reactions, or at very high pressures and temperatures. For example, some

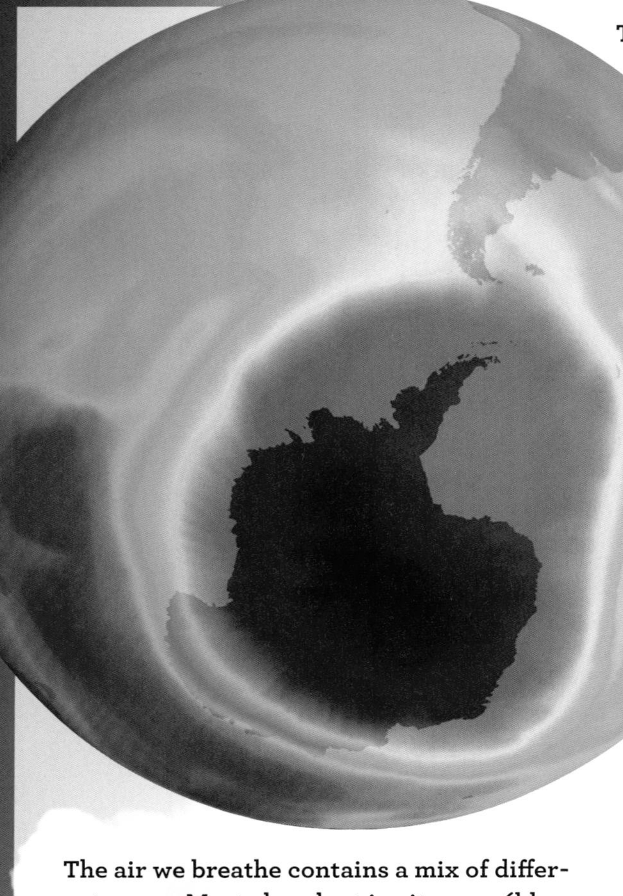

The depleted ozone layer over Antarctica is shaded purple. Ozone (O_3) molecules consist of three oxygen atoms, which break down to form normal two-atom molecules.

molecules in the atmosphere are broken up into their atoms in the intense heat of a lightning discharge. The separate atoms produced soon react together and recombine to form the same molecules again.

THE MOVEMENT OF MOLECULES

Molecules are always moving. In a solid such as a rock, each molecule is constantly vibrating. However, the molecules do not move away from their "home" position, so the solid material does not easily change its size and shape. A rock does not change its shape if you push it with a finger (but if you hit the rock with a hammer, it may break up).

In liquids the molecules also vibrate, but they have more freedom to move around than the molecules in a solid. However, they are packed quite close together (as in a solid), and they also tend to stay in contact with neighboring molecules. The molecules move around in a loose group, passing each other and sliding past other molecules in the liquid. This is why liquid moves around in its container: juice, for example, takes the shape of the cup, jug, or whatever container it is poured into.

The air we breathe contains a mix of different gases. Most abundant is nitrogen (blue atoms), followed by oxygen (red). The air's other gases, including carbon dioxide, argon, krypton, and the other noble gases, make up just 1 percent in total.

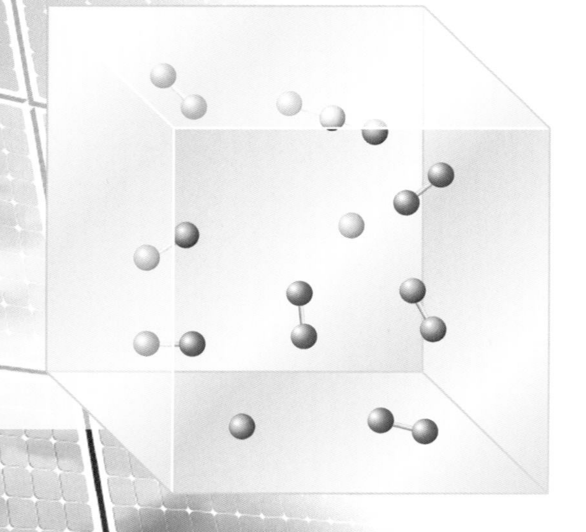

Molecular Size

Molecules can vary greatly in size. Most are tiny and are made up of just a few atoms. But some molecules in living things are much, much bigger. A single molecule of DNA or RNA (the genetic materials found in the centers of cells) can consist of millions of atoms. Stretched out, a molecule of DNA from a human cell would be about 2 inches (5 cm) long! Many protein molecules are also large, containing hundreds of atoms. A single crystal of ordinary salt, sodium chloride, can also be regarded as one huge molecule. Its atoms do not form small groups, but instead link with their neighbors to form a huge, regularly spaced network of atoms called a crystal lattice.

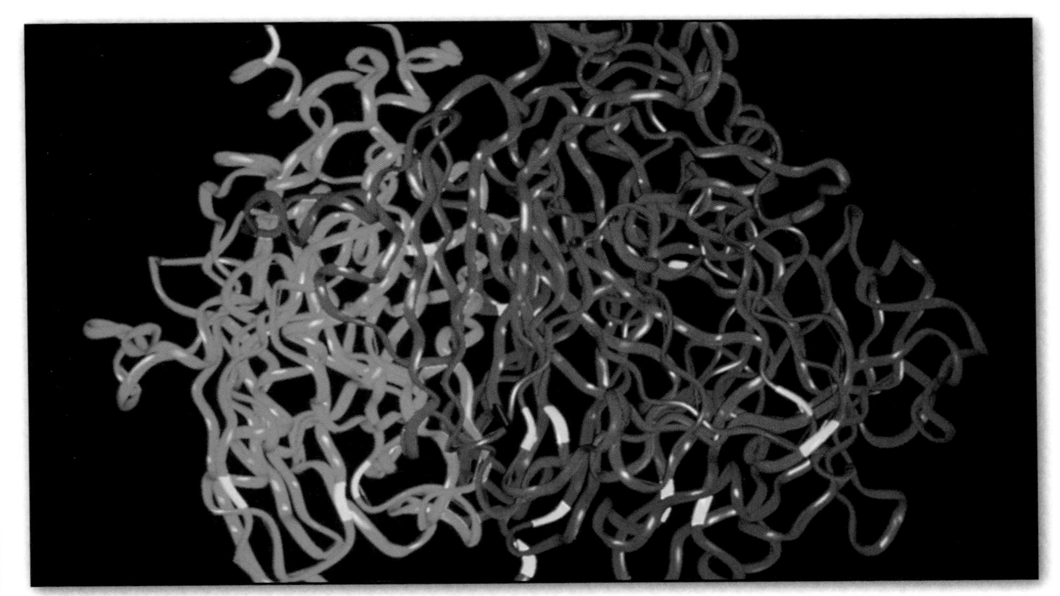

This huge molecule is the protein factor VIII, which plays an important role in helping blood to clot and stop bleeding.

In a gas, the molecules are far apart and can fly around. They collide with each other and bounce off the walls of the container. The molecules not only move around but also rotate and vibrate.

In air at normal temperatures—that is, about 68 degrees Fahrenheit (68ºF; 20ºC)—most of the molecules of oxygen and nitrogen move at approximately 1,500 feet per second (450 m/s), which is faster than the speed of sound. At any given moment, some molecules are traveling much faster than this, and some are moving much more slowly.

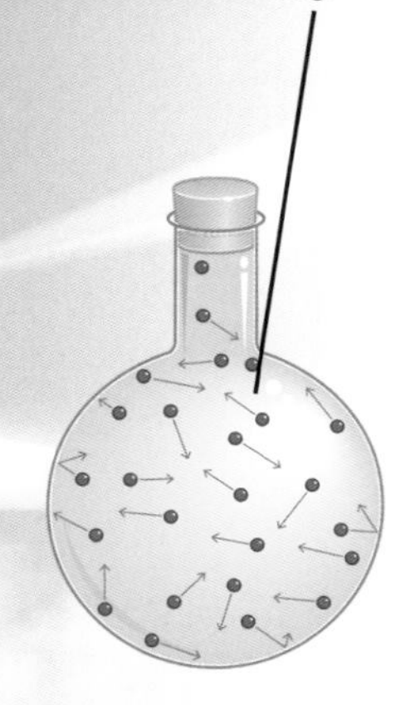

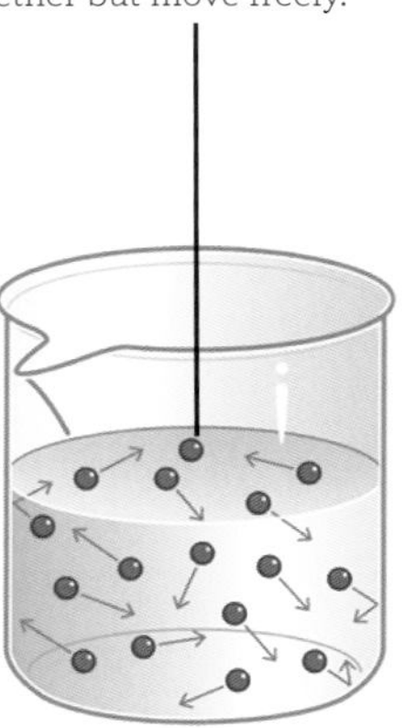

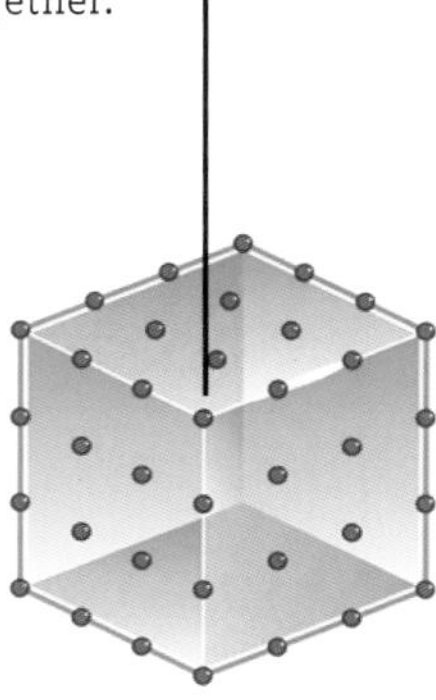

In the three main states of matter, the particles have different spacings and varying amounts of movement.

In a liquid, the particles can slide over each other producing a flowing mass, as in Niagara Falls.

The idea that matter consists of myriad tiny particles in motion is called the kinetic theory of matter. In everyday speech, the word "theory" is sometimes used to mean an idea that has not been tried in practice. However, in science it means something quite different. A theory is a wide-ranging and detailed description that closely matches the known facts. There is nothing uncertain about the kinetic theory: it is well confirmed by a great many experiments.

HOW GASES BEHAVE

The kinetic theory assumes that gas molecules are like tiny balls, bouncing around inside a container. This idea gave the theory its first success by explaining many aspects of how gases behave.

First, when a molecule hits a wall of the container, it bounces back with unchanged speed and unchanged kinetic energy. At the same time it pushes against the wall. This push is the pressure that we know a gas exerts on its container.

Also, if the gas is squeezed into a smaller space, the molecules bounce off the walls more often because they take less time moving from one wall to another. That explains why, as the gas is compressed and its volume decreases, the pressure increases.

If the molecules of a gas are speeded up, the pressure increases because the molecules exert a greater force on the

A military aircraft flies faster than the speed of sound. Molecules in the air around us are traveling at such high speeds all the time.

This hygrogen atom has been magnified to appear this size.

TAKING A CLOSER LOOK AT

ATOMIC SIZES

A typical size for an atom is 12 billionths of an inch (30 billionths of a centimeter) across. Many billions of atoms could fit into the period at the end of this sentence. However, atoms have a range of sizes. The largest atom, cesium, is about 20 billionths of an inch (50 billionths of a centimeter) across; the smallest, hydrogen, is 3.5 billionths of an inch (9 billionths of a centimeter) across.

walls when they bounce off them. The kinetic theory tells us that the average speed of the molecules increases as the temperature of the gas increases. Thus, that is why the pressure of a gas increases with temperature.

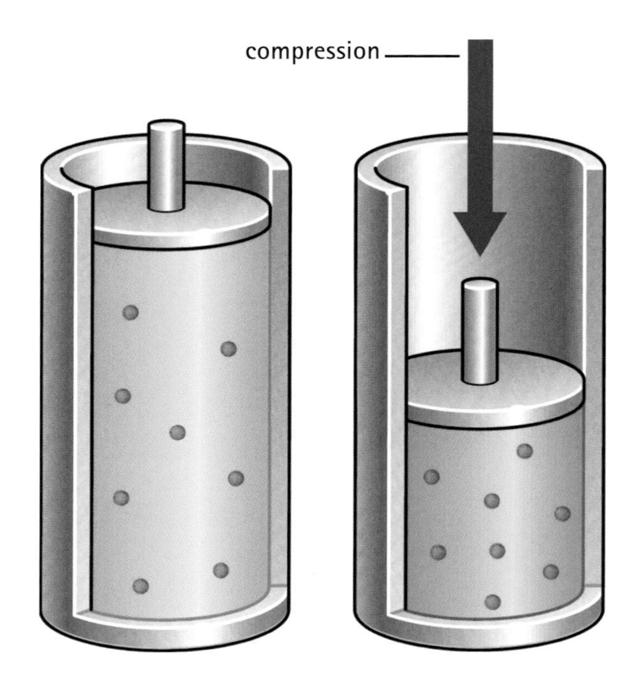

When some gas is compressed, its volume decreases but its pressure increases, as there are more collisions with every part of the container.

BELOW: The kinetic energy of gas molecules has several components. As well as translational motion (movement from one place to another), there are the twists and turns of the molecules themselves.

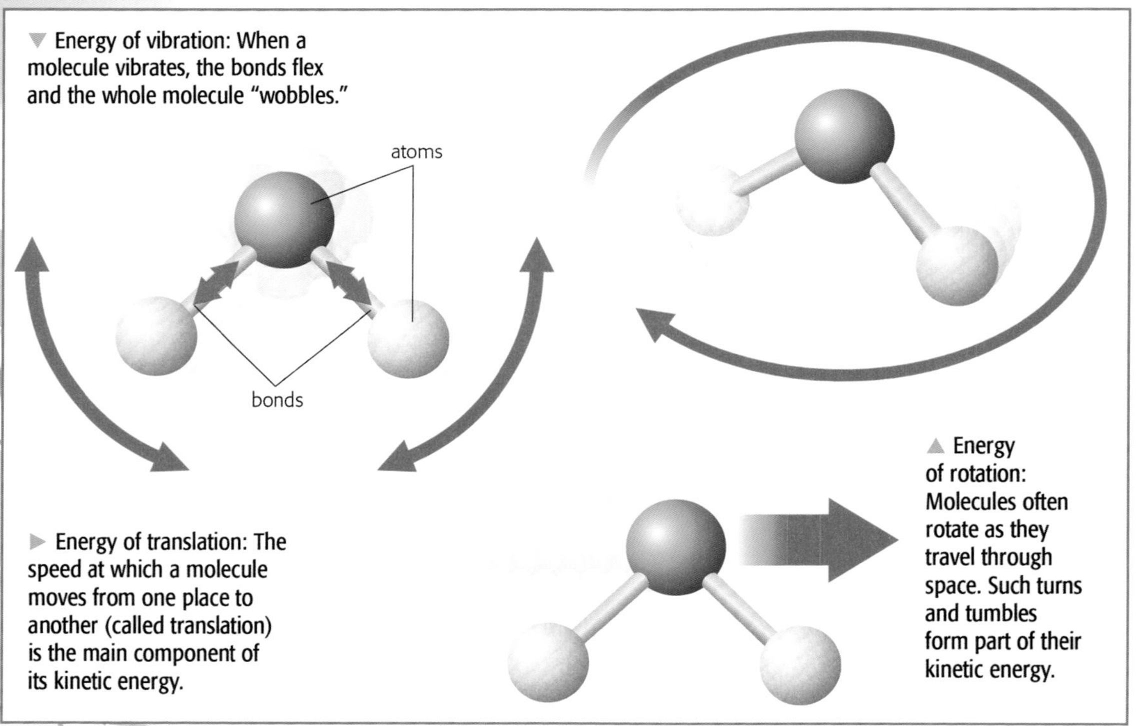

▼ Energy of vibration: When a molecule vibrates, the bonds flex and the whole molecule "wobbles."

▶ Energy of translation: The speed at which a molecule moves from one place to another (called translation) is the main component of its kinetic energy.

▲ Energy of rotation: Molecules often rotate as they travel through space. Such turns and tumbles form part of their kinetic energy.

KEY DEFINITIONS

- **Energy:** The ability to cause a change in something by heating it up, altering its shape, or making it move.
- **Kinetic energy:** The energy of movement.
- **Kinetic theory:** The study of heat flow and other processes in terms of the motion of the atoms and molecules involved.
- **Molecule:** The smallest particle of a chemical substance that can exist on its own. Molecules consist of atoms bonded together.

These relationships between volume, temperature, and pressure in a gas are called the gas laws. In fact, these laws only hold perfectly true for an "ideal" gas, which has these characteristics:

- The volume of the molecules is so small compared with the space around them that it can be ignored.

PIONEERS OF KINETIC THEORY

The great 17th-century English scientist Isaac Newton (1642–1727) believed that gases behave the way they do because their particles repel each other (push each other apart), even when they are not in contact. In 1738, the Dutch-born Swiss scientist Daniel Bernoulli (1700–1782) instead suggested, correctly, that the pressure of gases is due only to the collisions of their particles with the walls of the container. This important idea was largely ignored until English scientist John Herapath (1790–1868) revived it. Even then, scientists generally did not accept the kinetic theory until the experiments of English physicist James Prescott Joule (1818–1889) and the detailed mathematical work of Scottish physicist James Clerk Maxwell (1831–1879).

Portrait of the scientist Daniel Bernoulli, who did extensive work on fluids (liquids and gases).

- When the molecules collide with the walls of the container, they bounce off without losing any energy. So the total kinetic energy of the gas never decreases.

- The molecules are not attracted to each other or to the container walls.

Real gases are not exactly like this, but they do behave very similarly to ideal gases over a wide range of temperatures and pressures, and so the gas laws do apply to gases in practice. As well as explaining how gases behave, the idea of moving molecules also explains a great deal about solids and liquids, and about chemical reactions.

MOLECULAR MOTION

The constant motion of molecules is what we know as temperature: The hotter a piece of matter is, the faster its molecules move. As matter cools, its molecules move more and more slowly.

When two objects at different temperatures are in contact, the warmer one cools down and the cooler one warms up. So the molecules in the cooler object will start to move faster, and those in the warmer object will slow down. Kinetic energy is transferred from the faster-moving molecules of the hotter object to the slower-moving molecules of the cooler one.

This transfer of kinetic energy is called a flow of heat. A flow of heat is

Biological samples can be quickly cooled to very low temperatures by immersing them in liquid nitrogen, which has a temperature of around −328ºF (−200ºC).

Ice is carved into these curvy shapes by flowing water, which has a slightly higher temperature and so melts the ice at its edges.

not only caused by a change in temperature, but it also produces a change in temperature (the hot object gets cooler and the cool object gets warmer). However, as we know from the gas laws, heat flow can also cause other sorts of changes. For example, if gas in a closed container is heated, the gas will get warmer and its pressure will also increase. But if the gas is free to expand outside the container, heating will cause expansion rather than increase the pressure. So heat can make a gas expand or increase its pressure, as well as raise its temperature.

A flow of heat can also cause matter to change its state—that is, whether it is a solid, liquid, or a gas. When an ice cube floating in a glass of warm soda melts, heat flows into the ice from the soda. While the ice is still below its melting point of 32°F (0°C), the heat warms the ice up. When the ice reaches its melting point, it begins to turn into water. As the ice is melting, its temperature stays at 32°F (0°C). At this point the flow of heat is causing a change of state from solid to liquid, not a temperature change. Similarly, when water is boiled, heat flows into it, turning it into steam. The liquid water and the steam stay at the same temperature, 212°F (100°C), until all the water has been turned into steam.

WAYS TO MEASURE TEMPERATURE

Chemists and other scientists need to measure temperature accurately. To do this, they use thermometers. One familiar type of thermometer uses a thin column of liquid mercury (or alcohol) inside a glass tube. The mercury expands when it gets warmer, and the end of the column moves along the tube, which is marked in units of temperature. Temperature can also be measured by its effect on electrical circuits.

A current generally flows in a wire more easily if the temperature of the wire is lower, and so measuring the resistance of a wire can provide a way of measuring temperature. Instruments called thermocouples also use wires to measure temperature. Because thermocouples are cheap and more robust than glass thermometers, they are often very convenient to use. Another device called a pyrometer uses the fact that very hot materials give out light of a characteristic color—for example, the glow of red-hot iron. Analyzing the color of light from a hot object can be used to reveal its temperature.

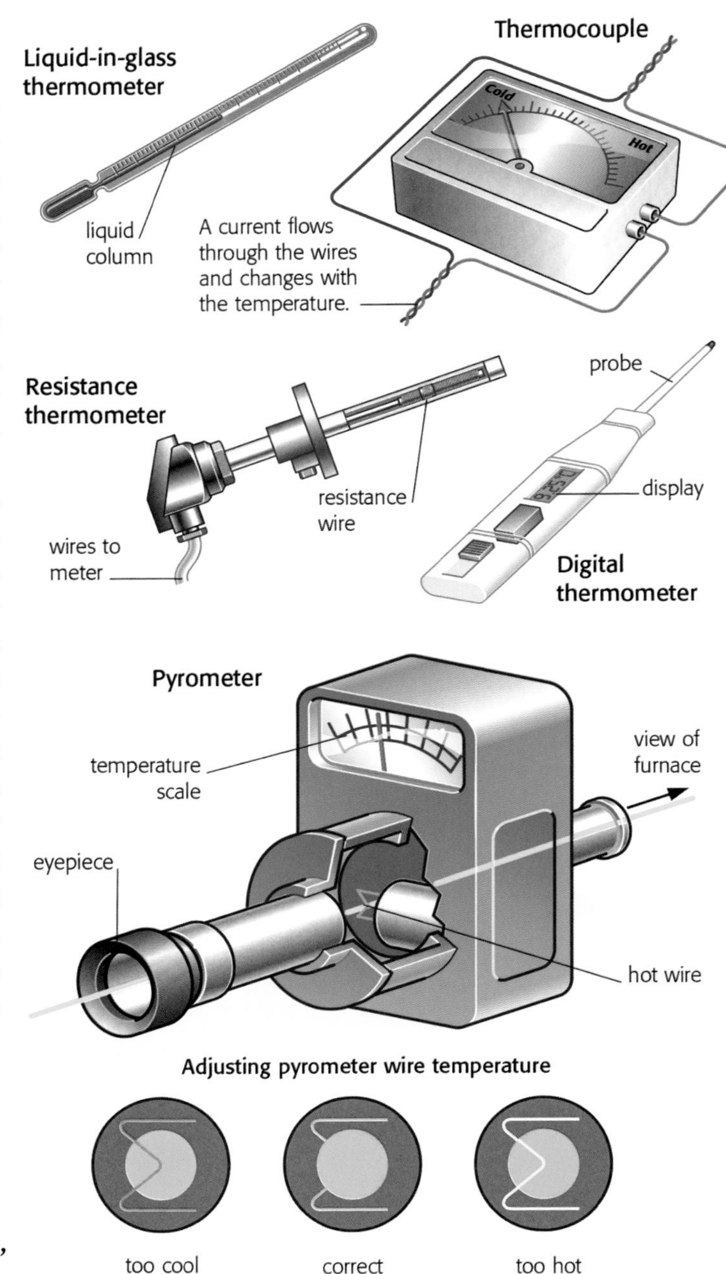

Many different instruments are used to measure temperature. Glass thermometers were traditionally most used, but electronic versions are now replacing these. Pyrometers use the light from hot, glowing objects to find temperatures. A measured current through a wire is adjusted until the wire's color matches that of the hot object.

CHEMICAL ENERGY AND BONDS

Matter contains energy because its molecules are moving, and because of the way in which the atoms that make up its molecules are joined.

When atoms link together in a chemical reaction, they are said to form chemical bonds. These bonds, or links, can be strong or weak, and there can be more than one bond between two atoms.

It takes energy to break a chemical bond because energy must be put in to pull the atoms apart. The atoms can be heated, or another atom can come into contact with them, forming new bonds as the old ones break. Once the atoms are separated, they still have this energy. So, if the atoms recombine and the bonds are re-formed, they give out the energy.

So forming chemical bonds means energy is given out, and breaking chemical bonds means energy must be put in. During chemical reactions, energy is taken in as bonds break and given out as bonds form. When a flask is heated in the laboratory to make a reaction happen, energy is taken in. Energy can also be taken in from radiation. For example, the chemicals in photographic film take in energy when exposed to light, which is a form of radiation. The chemicals react, producing a photographic image

An enormous amount of energy is needed to launch a rocket and propel it into space. This comes from the chemical energy stored in the rocket's fuel, which is released as the fuel burns.

The bangs, flashes, and sparkling colors in a firework display come from exothermic chemical reactions.

after the film has been chemically processed.

DIFFERENT TYPES OF REACTIONS

Chemical reactions that give out energy are called exothermic. The word comes from the Greek words *exo* meaning "outside" and *therme* meaning "heat." However, the energy given out can be in the form not only of heat, but also in the form of light, sound, movement, an electric current, and so on.

The burning of gas or wood is an exothermic reaction. Some of the heat given out travels away from the fire as heat radiation, and some of it goes into the ash, smoke, and gas produced. Some heat also goes into the unburned fuel, which keeps the reaction going (as it needs a high temperature to proceed).

In a similar way, reactions that take in energy are called endothermic, from the Greek words *endo* meaning "inside" and *therme* meaning "heat." When they are growing, plants absorb the energy of sunlight to build up their tissues. This process, which is called photosynthesis, is endothermic. Another endothermic reaction is the breakdown of calcium carbonate (limestone) into lime (calcium oxide) and carbon dioxide. This reaction needs a supply of heat to provide the energy required.

THE FIRST LAW OF THERMODYNAMICS

One of the most fundamental laws in the whole of science is the law of conservation of energy. It states that, in any chemical or physical process, the total energy of everything involved is the same afterward as it was before.

This means that, in a chemical reaction, the total sum of the kinetic energies of molecules, the energies of chemical bonds, and the energies of heat and light involved must be exactly the same after the reaction has taken place as they were before it.

The law of conservation of energy is basic to the science of thermodynamics, which deals with the relationships between heat and other forms of energy. That is why the law of conservation of energy is often called the first law of thermodynamics.

KEY DEFINITIONS

- **Bond:** A chemical link between atoms.
- **Endothermic reaction:** One in which the reacting substances take in heat.
- **Exothermic reaction:** One in which the reacting substances give out heat.
- **Heat:** Energy that flows as a result of a temperature difference.
- **Products:** The substances produced in a chemical reaction.
- **Reactants:** The substances that react together in a chemical reaction.
- **State of matter:** One of the forms that a substance can take, depending on its temperature and pressure. In everyday life we see three states of matter: solids, liquids, and gases.
- **Thermodynamics:** The study of how heat and other forms of energy are converted into each other.

Corn is a fast-growing crop. It uses the energy in sunlight to build molecules of the sugar glucose from water and oxygen. This endothermic reaction could not happen without the plant's complex chemical apparatus.

CHAPTER TWO

HEAT AND CHEMICAL REACTIONS

We obtain energy from the food we eat. The more calories a food has, the more energy it produces. Chocolate contains many calories.

Substances change during chemical reactions, and so does the amount of energy they store. These changes provide the energy that is given out in reactions.

Chemists often need to study the heat taken in or given out in chemical reactions. The amount of heat taken in or given out is called the heat of reaction. To measure heats of reaction, people use devices called calorimeters.

CALORIMETERS

There are various types of calorimeters. One type is called a bomb calorimeter because the reaction takes place inside a strong casing like that around a bomb. The casing must be able to withstand the high pressures that sometimes develop during a reaction. This casing is surrounded by water. The reaction that takes place inside either gives out heat, warming the water, or it takes in heat from the water, cooling it down. The experimenter measures the temperature of the water before and after the reaction takes place. From the temperature change the amount of heat that has flowed

COOL TOOLS

BOMB CALORIMETERS

In a bomb calorimeter, a substance is burned in oxygen to find out how much heat is given out. The substance is weighed and placed in the "bomb," and oxygen is pumped in under pressure. The reaction is started by an electrical spark. A stirrer is used to equalize the temperature throughout the water, while the insulated walls of the calorimeter prevent heat from escaping. The change in the water's temperature is measured using a thermometer.

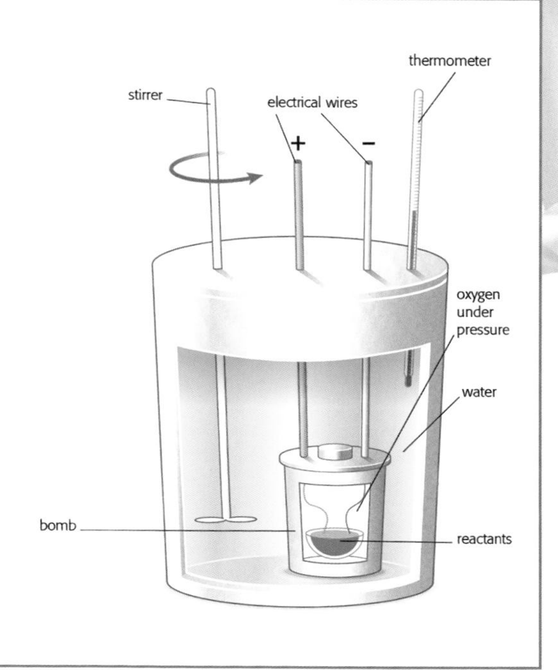

Nutrition Facts

Serving Size 1/2 cup (39g) dry
Servings Per Container about 13

Amount Per Serving	Cereal	Cereal with 1/2 cup Skim Milk
Calories	140	190
Calories from Fat	25	30
	% Daily Value**	
Total Fat 3g*	5%	5%
Saturated Fat 0.5g	3%	3%
Cholesterol 0mg	0%	0%
Sodium 0mg	0%	3%
Total Carbohydrate 26g	9%	11%
Dietary Fiber 4g	16%	16%
Soluble Fiber 2g		
Sugars 0g		
Protein 5g		
Vitamin A	0%	6%

in or out is calculated: the greater the change, the greater is the heat of reaction.

Another type of calorimeter, called a flame calorimeter, can be used for reactions that involve burning. Other types of calorimeters can be used for other types of reactions—for example, mixing an acid and a base so that they neutralize each other.

HEAT AND EXPANSION

Imagine a reaction taking place in a closed container (such as a calorimeter) so that the volume of all the reactants is

The labels on food packaging provide information on calorie value—that is, how much energy the food will yield when it is used in chemical reactions in the body.

COOL TOOLS

CALORIMETERS MADE OF COFFEE CUPS

A simple calorimeter for heat experiments can be made from two expanded polystyrene cups nested together, with a lid. Expanded polystyrene is a good heat insulator (heat does not easily flow through it), which is why cups made of this material are good for keeping coffee hot or soda cold. The experimenter places the experimental substances in the cups and mixes them together. These could be weighed quantities of reactants, or a solid and a liquid at different temperatures, depending on the experiment. He or she then reads the temperature using the thermometer pushed through the lid.

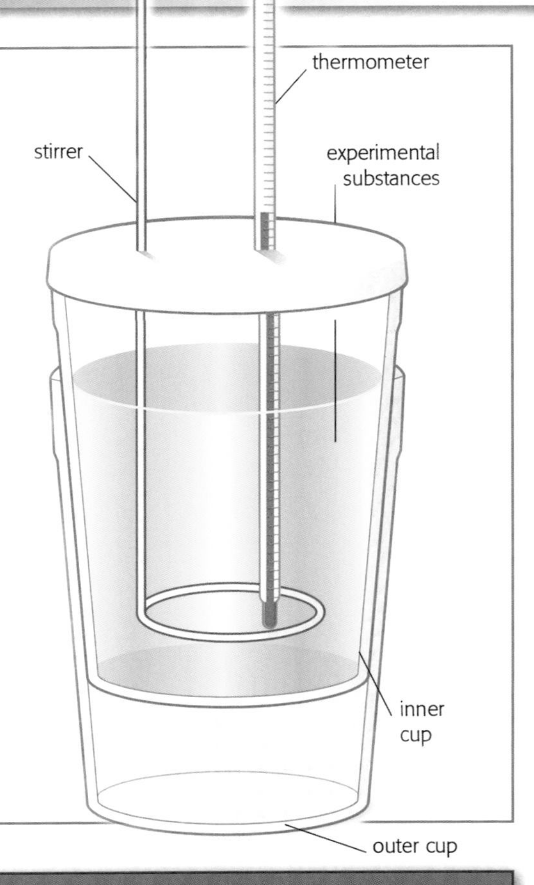

For heat experiments where an approximate result is sufficient, such as in middle-school investigations, a calorimeter can be made easily from polystyrene cups.

constant. The reaction gives out gas as a product, together with energy. Some of the energy is in the form of heat, so the reaction products rise in temperature. The rest of the energy produced is stored in the chemical bonds of the products.

If the same reaction takes place instead in an open container, the products will heat up slightly less than in the closed container. So, what has happened to the energy that has "disappeared?" The gas produced has expanded, pushing against the pressure of the surrounding air and using up energy as it does so. Less energy therefore remains to be turned into heat. As the heated gases expand and push back the air, they give a

TAKING A CLOSER LOOK AT

HEAT CAPACITY

To calculate the heat given out or taken in by a reaction, chemists need to know how much heat the calorimeter itself gives out or takes in. This means the experimenter first needs to find out what the "heat capacity" of the calorimeter is. The heat capacity of an object is the amount of heat needed to change its temperature by 1 degree Celsius (1°C; 1.8°F). The calorimeter's heat capacity is multiplied by the temperature change in the experiment to give the amount of heat flowing into or out of the apparatus.

Scientists use computers to analyze the large amount of data provided by modern calorimeters.

little bit of energy to the air molecules, so no energy has been lost. (The solids and liquids involved change volume by only a tiny amount compared with the gases, so the energy they lose through expansion would not be noticeable.)

WORK

When heated gases expand, they do what scientists call work. Work is done whenever a force moves something. The amount of work is defined as the force multiplied by the distance moved in the direction of the force. So, suppose you lift a heavy book off the floor onto a desk. You apply a force to lift the book (against the force of gravity) through the distance from the floor to the desktop. The energy to do this work was provided by chemical reactions in your muscles. Some chemical energy in your body has been used up in doing work.

KEY DEFINITIONS

- **Calorimeter:** Apparatus for accurately measuring heat given out or taken in.
- **Heat capacity:** The amount of heat required to change the temperature of an object by 1°C (1.8°F).
- **Specific heat capacity**: The amount of heat required to change the temperature of a specified amount of a substance by 1°C (1.8°F).

When we bowl or throw a ball, chemical reactions in our muscles convert energy stored in the body into energy of movement.

TAKING A CLOSER LOOK AT

DIFFERENT TYPES OF SYSTEMS

In chemistry, the word "system" means the particular things that the chemist is dealing with at the time. In a chemical reaction it includes the reactants, the products, any energy produced, and often the container they are in. There are different types of systems:

- A system is open if matter and energy can flow between it and its surroundings. An open flask containing reactants is an example of an open system.

- A system is closed if no matter can be exchanged between the system and its surroundings, but energy can still be exchanged. A closed flask is an example of a closed system because heat can flow through its walls but matter cannot.

- A system is isolated if neither energy nor matter can flow into or out of it. No real system is perfectly isolated. The closest thing to an isolated system that we normally encounter is a vacuum flask *(left and below)*, although even this lets heat out (or in) very slowly.

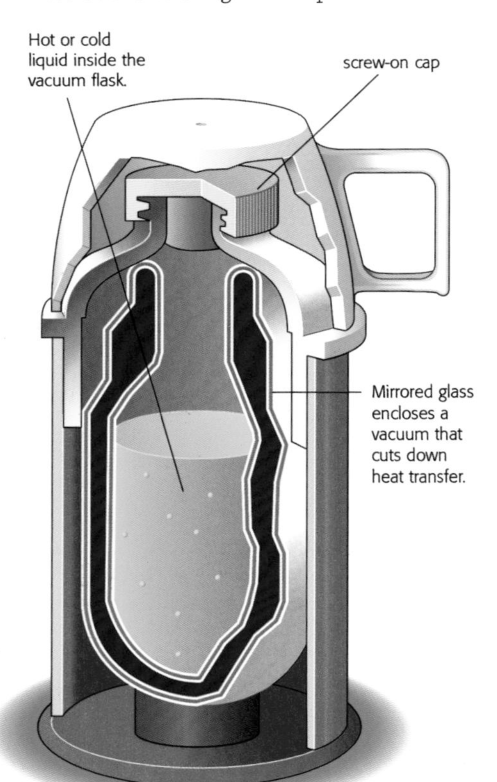

Work can be done in producing other kinds of energy. When you roll a bowling ball, the force of your hand moves the ball through a certain distance before you let it go. Your body does work, and gives the ball kinetic energy.

UNDERSTANDING POTENTIAL ENERGY

Kinetic energy is one form of energy and heat is another form. There is also another form of energy called potential energy. Things have potential energy because of their position or the arrangement of their parts. When you lift a book onto a desk, you give it potential energy. (In fact, the potential energy gained is equal to the work done in lifting the book.) Potential energy can be converted into other forms of energy: for example, if the book were to fall off the desk, some of the potential energy would be converted into kinetic energy (energy of motion). A stretched spring also has potential energy: when released, it jumps back to its original length and the potential energy is converted into kinetic energy.

Lifting weights in a gym is something many people do in their leisure time, but in scientific terms they are doing work. Every time we move an object against a force—such as moving a barbell upward against the force of gravity—we are doing work.

Chemical energy is also a kind of potential energy. The energy stored in chemical bonds can be converted into heat, light, sound, or motion when a chemical reaction takes place. Another important type of potential energy is electrical potential energy, which makes an electric current flow.

LET'S TRY THIS

HEAT LOSS

Materials: **a vacuum flask; a jug (preferably with a lid, or a cloth to cover the top instead); and a thermometer**

For this experiment, use hot water from the faucet. (This is hot enough: do not use boiling water.) First, warm the flask and jug by filling both with hot water and leaving for a few minutes. Then pour out the water.

Next, fill the vacuum flask, pour that water into the jug and put the lid on. Refill the vacuum flask and put its cap on tightly. This ensures the jug contains the same amount of water as the flask so that you can make a fair comparison between them.

The jug should be filled using the vacuum flask. That way the flask and the jug will contain the same amount of water.

Measure the temperature of the water in both the jug and the flask at regular intervals using a thermometer.

Measure and write down the temperature of the water in the two containers after 20 minutes, 40 minutes, and 1 hour. Each time you measure the temperature, keep the thermometer in the water for 30 seconds to give the thermometer time to come to the right temperature. Quickly re-stopper the flask and cover the jug after each temperature measurement.

Label your tables of measurements "isolated system" for the flask and "closed system" for the jug. You can also plot graphs of the temperatures against the time passed.

Finally, repeat the experiment with the jug, again filling it from the vacuum flask. This time, leave the lid off the jug. Label these results "open system." Again, you can plot the graph.

You should have found that the system that lost heat fastest was the jug without a lid. This is an open system: it can exchange energy and matter with its surroundings. Some of the water molecules escaped into the air, taking heat energy with them.

The jug with a lid on is a closed system; the hot water all stayed inside. It should have lost heat not as fast as the open jug. The vacuum flask is close to being an isolated system. It is designed to lose heat only very slowly, and its tight stopper prevents water vapor from being lost. You should have found that its temperature went down by the smallest amount over the times measured.

In many processes, potential energy decreases as it is converted into other sorts of energy. As a general rule, systems tend to lose potential energy.

For example, when an oil drum lying at the top of a hill rolls down the hill, it has less potential energy at the bottom of the hill than it does at the top. Similarly, a chemical reaction is likely to happen if the products have less chemical energy than the reactants.

Rusting is a reaction that releases a lot of energy. The reaction product is hydrated iron oxide, made up of iron, oxygen, and water. This has much less energy than the reactants that go into it. Although the rusting process

The elastic band in this catapult stores potential energy when it is pulled tight. When let go, this turns into kinetic energy, sending an object flying.

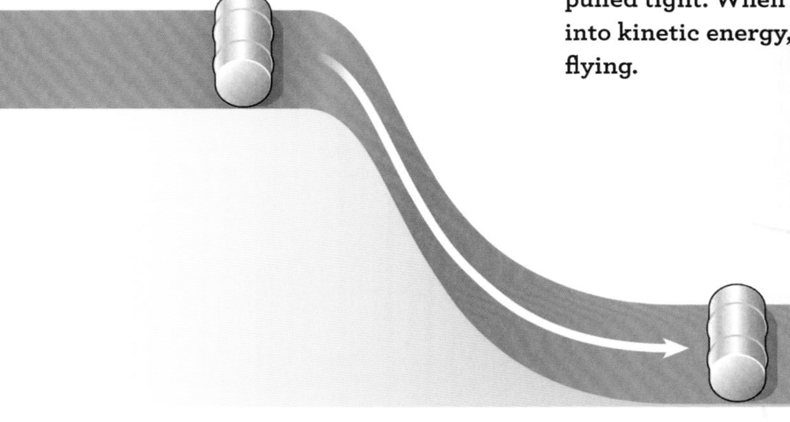

A rolling object has less potential energy at the foot of a hill than it had at the top. Many chemical reactions also lose energy as they progress.

LOOKING AT CHEMISTRY IN ACTION

THE CHEMISTRY OF CAR BATTERIES

Chemical reactions can release electrical energy or store it up because electrical energy can flow in or out of the substances involved in a chemical reaction. When a car is at rest, the car battery uses chemical reactions to deliver electrical energy to the lights and the radio, where it is converted into light and sound. When the car is moving, a generator connected to the engine drives an electric current through the battery in the opposite direction. This reverses the chemical reactions in the battery and stores energy for later use.

Cars have rechargeable batteries. Energy used to power the lights and radio when the car is at rest is replaced once the car is moving.

gives out energy and needs no heat to begin, it occurs only slowly. Comparing it with the oil drum rolling downhill, it is as if the oil drum were rolling through mud.

HELPING REACTIONS ALONG

Even if a mixture of chemical substances can lose energy by changing in some process, that process may not happen unaided. In the same way, if an oil drum at the top of the hill is

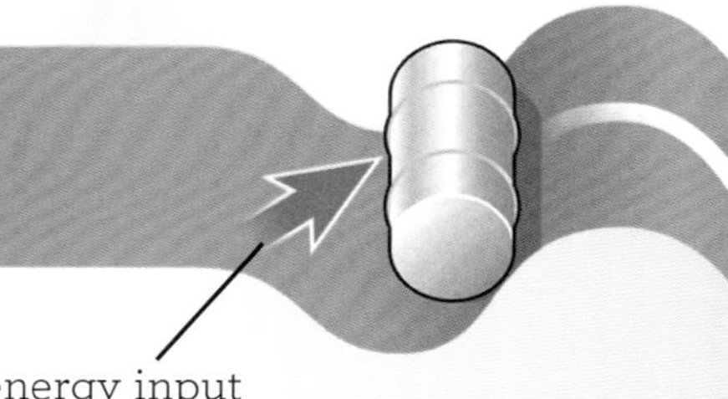

This rolling object needs to climb out of the dip before it can roll down the hill, so it will need energy to start it off. In a similar way, many chemical reactions need to take in energy at the start, even though much more energy is given out later as the reaction proceeds.

A spark or a flame needs to be applied to logs to get them to start burning.

standing upright, it will need to be toppled over to start it rolling. If it is on its side but is lying in a shallow hole, it will need a shove to lift it out of the hole and start it moving.

Burning paper is another chemical reaction that, like rusting, releases a great deal of energy. Again, the energy of the products is less than that of the reactants, with the energy difference given out as heat. But the paper does not burst into flames spontaneously (by itself). It needs some help from a lighted match or another source of heat to start it off. This is rather like an oil drum lying in a hollow, which needs a jolt to get it going.

HOW ENERGY IS ABSORBED

Chemical reactions in which the products have higher energy than the reactants have to take in energy to go ahead. The cold packs found in first-aid kits use reactions like this. When the pack is used, it rapidly becomes very cold and can be pressed against wounds to ease pain and reduce swellings. One type of cold pack consists of a strong outer bag containing powdered ammonium nitrate and an inner bag made of a weaker plastic and containing water. If the outer bag is struck or squeezed hard,

TAKING A CLOSER LOOK AT

HOW HEAT IS ABSORBED

Most chemical reactions give out heat, but some processes take heat in. These are some examples:

- Ice cubes melt in a glass of water, taking in heat from the surrounding liquid.
- Water in a dish evaporates in a few days. Molecules of the liquid escape into the air as the liquid absorbs heat from its surroundings. Volatile (easily evaporated) liquids such as ethanol (an alcohol) will disappear even sooner than water.
- Ordinary salt dissolves in water, absorbing heat. Many other substances also take in heat as they dissolve.

TAKING A CLOSER LOOK AT

HOW ELECTROMAGNETIC RADIATION WORKS

Energy can enter or leave a chemical reaction as radiation in various forms. Light and heat radiation form part of the electromagnetic spectrum—a vastly extended version of the visible light spectrum, which we see in a rainbow. The electromagnetic spectrum includes all the different forms of electromagnetic radiation arranged according to their wavelength (the distance from one wave's peak or trough to the next wave's peak or trough). At one end there are radio waves, which can have wavelengths more than 1 mile (1.5 km) long; at the other end are gamma rays—very high-energy waves given out in nuclear reactions. The main types of electromagnetic radiation that are important in chemical reactions are:

- Visible light, which is given out in many chemical reactions including exploding fireworks. The light given off by glow-worms and fireflies also comes from chemical reactions in their bodies. Light can cause chemical reactions, as well as be emitted in reactions.

- Infrared radiation ("heat radiation"), which has longer wavelengths than visible light and, like light, can travel through space (unlike other forms of heating).

- Ultraviolet radiation, which has shorter wavelengths than visible light and which causes the burning and tanning reactions in our skin.

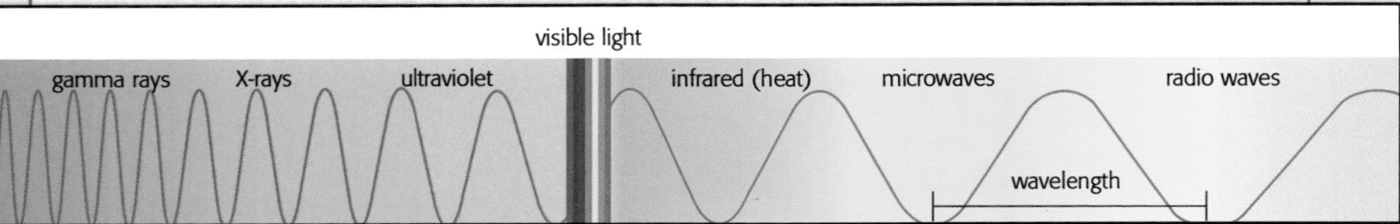

In the electromagnetic spectrum, gamma rays have the highest energies and shortest wavelengths (about a hundred billionths of a millimeter long). Radio waves have wavelengths about a hundred million million times larger.

the inner bag bursts and the water and ammonium nitrate mix. The ammonium nitrate dissolves in the water, absorbing heat so strongly that the temperature of the liquid can drop to 32°F (0°C).

In this endothermic reaction, the products of the dissolving process have higher energy in their chemical bonds than the reactants. The extra energy is taken in from the reactants and their surroundings, lowering their temperature. Although the products have gained energy compared with the reactants, no energy has been created or destroyed overall. The total amount of energy stays the same, in line with the first law of thermodynamics. The energy taken in from the surroundings is stored in the products as chemical energy within the bonds.

ENTHALPY AND INTERNAL ENERGY

The total energy within a piece of matter is called its internal energy. This is made up of the total kinetic energy of all the particles in the matter (due to their movements), plus the potential energy stored in the chemical bonds of the substances that make it up. If heat is given out or taken in during a reaction, the internal energy of the substances will decrease or increase accordingly.

Chemists use the term "enthalpy change" for changes in internal energy that take place during reactions. If the pressure is constant (such as when the reaction takes place at normal air pressure), the enthalpy change in a reaction is equal to the change in internal energy between the reactants and the products.

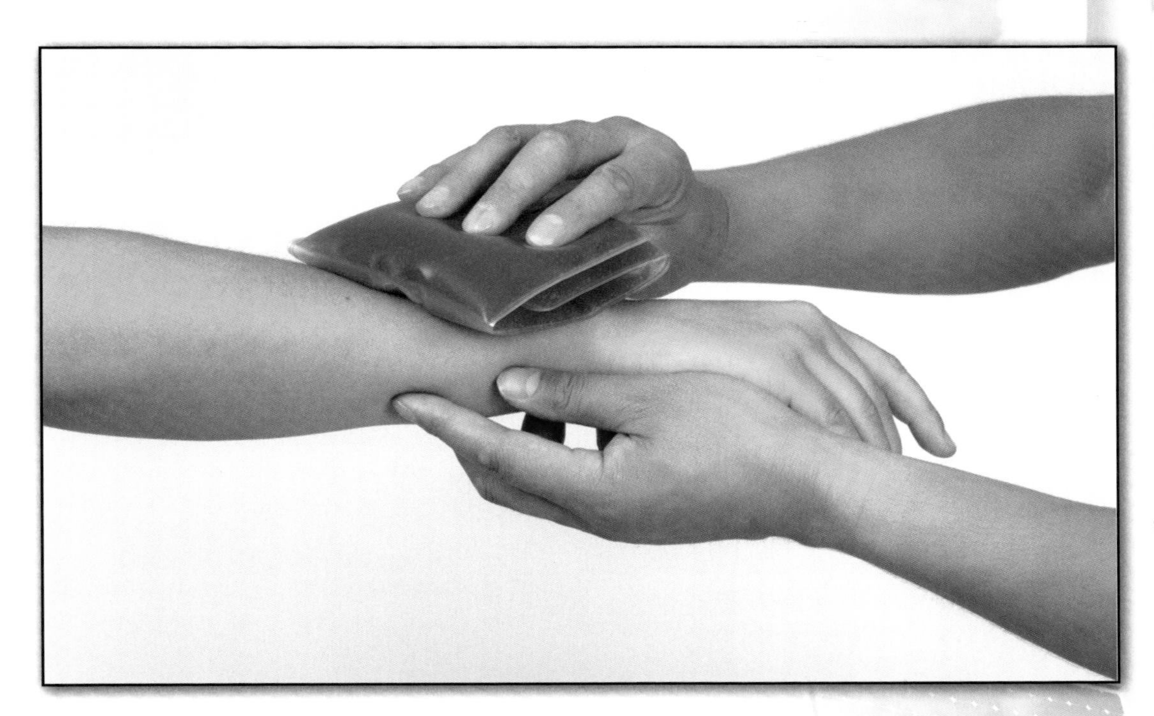

A sprained wrist is soothed by a cold pack applied to it. Inside the pack, an endothermic reaction takes place that draws heat in.

When water is heated, its energy increases and so its change in enthalpy is positive.

KEY DEFINITIONS

- **Electromagnetic radiation:** Radiation forming part of the electromagnetic spectrum, such as light, heat radiation, and ultraviolet light.
- **Enthalpy change:** At constant pressure, the change in internal energy that occurs during any process.
- **Internal energy:** The total kinetic energy of all the particles in a system, plus all the chemical energy.
- **Potential energy:** Energy that something has because of the way it is positioned or how its parts are arranged.
- **Spontaneous reaction:** A reaction that happens by itself, without needing something from outside to start it off.
- **Wavelength:** The distance measured from the peak or trough of one wave to the peak or trough of the next.

Enthalpy is measured in the same units as energy: joules or calories. If a reaction is exothermic (that is, energy is given out), the enthalpy change is a negative amount because the energy of the products is less than that of the reactants. Similarly, for endothermic reactions, enthalpy changes are positive. For example, the enthalpy change when 1 gram of water evaporates is 2.26 kilojoules (a kilojoule is 1,000 joules), or 9.46 kilocalories. The reverse process (when 1 gram of water condenses from gas to liquid) has an enthalpy change of minus 2.26 kilojoules (minus 9.46 kilocalories).

UNDERSTANDING HESS'S LAW

Hess's law was formulated by German Henri Hess (1802–1850). The law states that the overall enthalpy change that accompanies a chemical reaction is independent of the route by which the reaction takes place. So, for example, if substance A can be converted into substance B by two different chemical processes, the total energy changes in each process must add up to the same amount. Hess's law is a version of the law of conservation of energy (or the first law of thermodynamics), since otherwise energy would need to be created or destroyed along the way.

CHAPTER THREE

WHAT ARE ENTROPY AND FREE ENERGY?

Energy drives reactions, but so does another factor called entropy. Although we are normally unaware of entropy, it also plays a part in everyday life.

What causes some chemical reactions to occur spontaneously (by themselves)? We have seen in the previous chapter that many reactions give out energy, but not all; and not all reactions that give out energy happen spontaneously. To explain which reactions occur and which do not, we need to look at another factor that is at work alongside energy, acting as a driving force in reactions. This important factor is called entropy.

Entropy is the driving force that makes evaporation happen, forming the mist and clouds in this lake landscape.

WHAT'S ENTROPY?

Entropy is related to disorder. To get an idea of what entropy is, imagine a gas-tight container is divided into halves by a partition, and that one half (call it compartment A) contains gas at ordinary atmospheric pressure, while the other half (compartment B) is completely empty. If a door in the partition is opened, what will happen?

We would expect the gas to instantly begin diffusing (spreading) from compartment A into compartment B. It will continue to do so until the gas fills the two halves of the container at equal pressure. In the same way, we do not expect all the air in a room to occasionally clump together in one corner and stay there.

After the door in the central partition is opened, all the molecules of the gas could wander into one half of the container or the other, just by chance. That would not violate any physical laws, such as the law of conservation of energy. It is simply very, very unlikely. Since the molecules are in constant movement, it is likely that the molecules will fill the space in the course of their wanderings. If instead we found gas crowding into one part of the container, we would think that some force was organizing the molecules. (Just as, if we saw people in a park gathering in one place, we would think something was happening to draw them to that area.)

Having all the molecules crowded together in one part of the available space is described as the gas being in an organized state. When they are spread out evenly through the available space, that is described as a disorganized state—or a "nothing special" state.

Entropy is a measure of disorder, so the entropy of the spread-out, disorganized gas is much higher than that of the crowded-together gas. The gas naturally

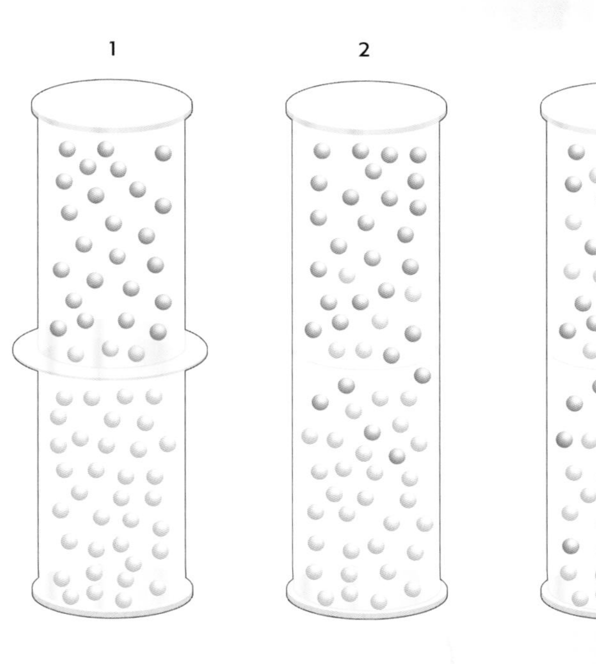

Two different gases are separately stored in a container (1). When the partition is taken away (2), the gases begin to diffuse into each other's space and entropy increases. Eventually, the gases become more evenly mixed throughout (3).

LET'S TRY THIS

WHAT ARE THE ODDS?

To get a feel for entropy, you can try calculating the probability of a small number of gas molecules all crowding into one half of a container divided into two compartments, A and B.

To make this simpler, imagine there are only four molecules in the container instead of the vast number that there really would be. If you checked their positions at a given moment, you might find three molecules in A and one in B, which you could write as AAAB; or you might find the first and third molecules in A and the second and fourth in B, which you could write as ABAB; and so on. These are different "microstates" of the gas.

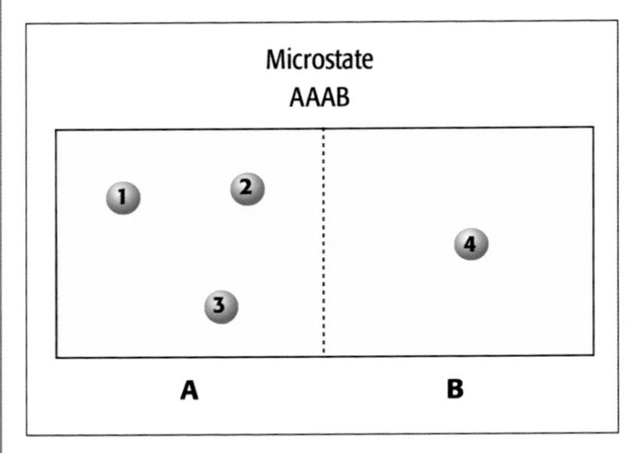

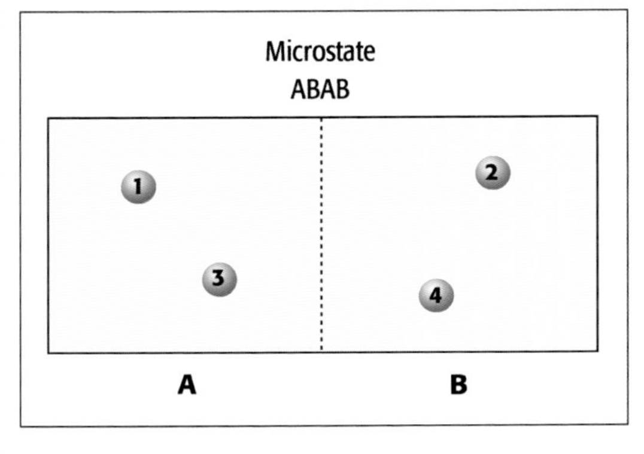

First, write down all the possible microstates of the four molecules in the two compartments.

You should find 16 microstates in all. These are: AAAA, AAAB, AABA, AABB, ABAA, ABAB, ABBA, ABBB, BAAA, BAAB, BABA, BABB, BBAA, BBAB, BBBA, and BBBB.

- Now count up, from among these 16 microstates:
- The number of microstates in which all molecules are in one compartment, A or B.
- The number of microstates in which there are three molecules in one compartment and one molecule in the other compartment.
- The number of microstates in which there are equal numbers of molecules in both compartments.

You should find that there is one microstate (AAAA) in which all the molecules are in A and one (BBBB) in which they are all in B. So there are 2 chances in 16, or 1 in 8, of all the molecules of the gas being in one compartment at any moment.

There are eight microstates in which there are three molecules in one compartment and one in the other;

and six microstates in which there are two molecules in each compartment. Adding these together gives 14 microstates (out of a total of 16) in which, at a given moment, the molecules are roughly evenly divided between the two compartments. That is a probability of 7 in 8—much higher than the 1 in 8 probability of all the molecules being in one compartment.

So, even for a very few molecules, finding the gas squeezed into one compartment is unlikely. The same thing holds true even more strongly for real gases, where there are trillions and trillions of molecules in even a tiny volume. Then, it is overwhelmingly likely that the gas will spread evenly through the available space.

tends to go from the highly organized state to the less organized state, so its entropy tends to increase. We find this, too, in everyday life. Shoelaces often come undone, but they never tie themselves into neat bows. It takes a special effort to tidy a bedroom, but no special effort to make it untidy. So, unlike with energy, the total entropy of the universe generally increases all the time (although in particular locations, it may decrease for a while).

ENTROPY COMPARED WITH ENERGY

Entropy increases in many processes. When a liquid evaporates, its molecules leave the liquid to move about in the space above it. They are less organized than when they were confined to the liquid, so the system's entropy increases.

When they are in the liquid, the molecules stay together because they attract each other. When a molecule leaves the liquid, each molecule does work by moving against these forces.

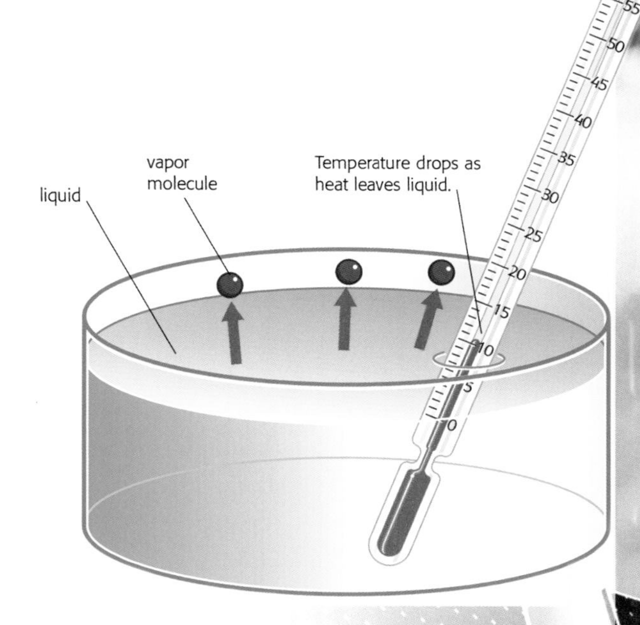

As we have all noticed, liquids left in an uncovered dish will eventually evaporate (become a vapor). That occurs because of entropy.

Hanging clothes outside on a warm, breezy day helps the water evaporate, making the clothes dry.

The molecules therefore need energy to evaporate. They take this energy from the liquid, cooling it down.

In some chemical reactions, there is more gas in the products formed than in the initial reactants (for example, when a solid and liquid are mixed and a gas is produced). In these reactions, the entropy almost always increases. The disorder of molecules when they are spread out within a gas is far greater than when they are confined to the much smaller volume of a solid or liquid.

TAKING A CLOSER LOOK AT

GASES AND VAPORS

In both vapors and gases, atoms and molecules move freely with a large amount of space between them. So are these two words for the same thing, or are gases and vapors actually different?

As a liquid is heated, the amount of vapor above it (called the vapor pressure) increases. When this becomes equal to the surrounding air pressure, the liquid boils and becomes a gas. Water normally boils at 212 degrees Fahrenheit (212°F; 100°C), since this is when its vapor pressure equals the air pressure. It becomes steam, which is a gas. Below this temperature it is water vapor, not steam.

The steam produced in a steam engine's boiler turns to water vapor when it reaches the air, producing the billowing white clouds we see.

Willard Gibbs (1839–1903), who put forward the important scientific idea now known as Gibbs free energy.

UNDERSTANDING GIBBS FREE ENERGY

The ideas of energy and entropy can be put together to understand exactly why some reactions happen spontaneously and others do not. But one extra idea is needed: that of Gibbs free energy, named for the U.S. scientist Willard Gibbs.

Gibbs discovered an equation that used the changes in enthalpy and entropy in a reaction to predict whether the reaction would occur. This equation makes the following statement:

CHANGE IN GIBBS FREE ENERGY = (CHANGE IN ENTHALPY) – (TEMPERATURE X CHANGE IN ENTROPY)

In other words, the change in entropy (multiplied by the temperature) has to be taken away from the enthalpy change in the reaction to give the change in Gibbs free energy. This is very important because it is only when the Gibbs free energy decreases (that is, the change in

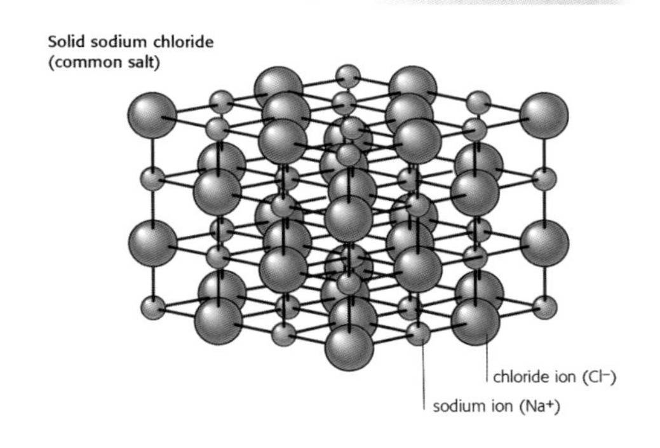

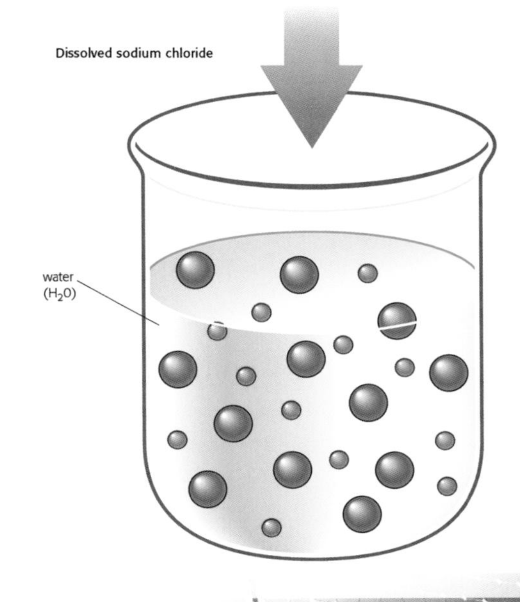

Gibbs free energy is negative) that a reaction will take place spontaneously. So, if the change in entropy is large enough, it can produce a negative change in Gibbs free energy overall, even if the enthalpy change is positive (as in an endothermic reaction). That is why some reactions that take in energy do occur spontaneously.

For example, in dissolving, the increase in entropy produces a negative change in Gibbs free energy, even though energy is taken in. In this case, the driving force of the entropy increase wins out over the braking effect of the enthalpy increase. Similarly, an exothermic reaction that produces a decrease in entropy may not happen spontaneously because the decrease in entropy cancels out the negative enthalpy change.

One form of the second law of thermodynamics tells us that heat always flows to a cooler body from a hotter one—which is why licking a Popsicle on a hot day is so refreshing.

HOW HEAT FLOWS

Heat diffuses (spreads) in much the same way that a gas does. If a hot body and a cold body come into contact, the molecular motions of the hot body spread to the molecules of the cooler body. This is because in a hot body, the average kinetic energy of the molecules is higher than in a cooler one. So when a molecule in the hot body collides with one in the cooler body, some kinetic energy will be passed from the hot body to the cooler one. In this way kinetic energy is gradually passed from the hotter body to the cooler one, which amounts to a flow of heat in the same direction. A flow of heat in the other direction (from the cooler body to the hotter one) is not forbidden by the first law of thermodynamics, as energy would not be created or destroyed. However, it is fantastically improbable, just as gas molecules failing to spread out into an empty space is so unlikely.

THE SECOND LAW OF THERMODYNAMICS

An isolated system is one that cannot exchange matter or energy with its surroundings. The second law of thermodynamics states that, in an isolated system, entropy will increase or stay the same; it will never decrease.

A vacuum flask is close to being an isolated system (although it slowly loses

In a refrigerator, electrical energy is used to keep the interior cool so that heat flows away from the warmer food placed inside.

heat energy to its surroundings). Imagine putting some hot water into the flask. At first, the water is hot and the flask itself is cooler, but then temperature differences start to disappear. The hot water cools a little as it warms the inner wall of the flask. Slight differences in temperature between the different parts of the liquid even out. As the temperature differences disappear, the system becomes less organized. Eventually the liquid and the inner wall of the flask are at the same temperature and the system has maximum entropy.

On a very large scale, the second law of thermodynamics applies to the universe as a whole. For example, a cluster of stars, perhaps circled by planets, is a relatively low-entropy system because it is highly ordered and organized compared with the mass of gas and dust from which it formed. But as the stars and planets developed, they gave out radiation energy. This spread out into space and warmed interstellar gas and dust. The increase in temperature raised the entropy in interstellar space, and increased the entropy of the universe itself.

Another form of the second law of thermodynamics refers to temperature rather than entropy.

In this form, the law states that heat never flows by itself from a cooler body to a hotter body. If it seems to do so, something else is driving the process. (This form of the second law looks very different from the entropy version, but in thermodynamic terms it is exactly equivalent.)

For example, in a refrigerator, heat is continually removed from the cool interior. This heat flows out from cooling pipes behind the refrigerator into the room, which of course is warmer than the inside of the fridge. But this flow of heat from cool to warm is powered by the fridge's electric motor.

A BRIEF HISTORY

MAXWELL'S DEMON

The 19th-century Scottish scientist James Clerk Maxwell famously posed a problem that suggested that the second law of thermodynamics could be broken. He imagined a tiny "demon" that could open and shut a trapdoor in the wall between two halves of an isolated gas container. The demon would allow fast-moving molecules coming from the left and slow-moving molecules coming from the right to pass through the trapdoor, and it would shut the door against other molecules. The result would be that hot gas would build up on the right side of the container and cold gas on the left. A scientist observing this would see gas originally at one temperature separating into hot and cold portions, with heat flowing from the cooler part of the container to the warmer part. The overall entropy would decrease, even though the system is completely closed.

Could we make tiny machines (or "nanobots") that could act like the demon and break the second law of thermodynamics in this way? In fact, we could not. The demon is a physical system, interacting with the gas molecules. Its own entropy increases as the entropy of the gas decreases. The entropy of the whole system, demon and gas combined, therefore increases—just as it should do according to the second law.

James Clerk Maxwell (1831—1879) was a mathematical physicist whose work revolutionized 19th-century science.

UNDERSTANDING ENTROPY IN OPEN SYSTEMS

Most real-life systems are open: they can exchange matter and energy with their surroundings. In an open system, entropy can decrease if the entropy of its surrounding increases at the same time.

For example, living things consist of structures that would be highly improbable if they came about by chance. Living things are, in fact, organized by the genes (units of inherited information) within cells. When an animal grows and develops, the process involves an enormous decrease in entropy. But this decrease is more than offset by the increase in entropy passing to the living creature's surroundings as it excretes waste matter and heat.

In a freezer, water is frozen and its entropy decreases as its molecules pass from the relatively disordered state of a liquid to the more ordered state of crystalline ice. But the entropy of the freezer plus its surroundings increases because waste heat is pumped out of the back of the machine and increases the entropy of the air there. In a similar way, when water freezes spontaneously in cold weather, its own entropy decreases; however, the entropy of the water plus its surroundings increases because of the heat given out in the freezing process.

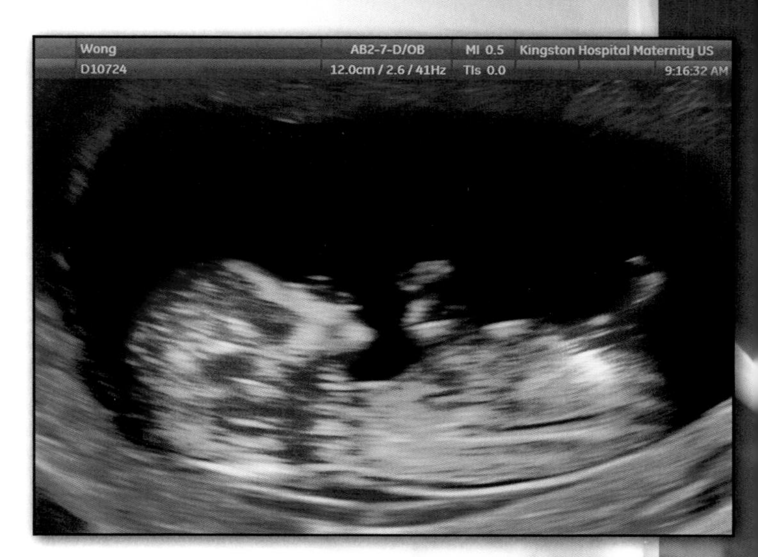

Hospitals use scans to monitor the way that a fetus is developing into a baby. This precisely ordered process involves a decrease in entropy.

As a crystal grows in nature, atoms are added in turn to a regularly spaced lattice, producing matter in its most organized form. The shapes in this uncut emerald reflect the arrangement of atoms in its lattice.

The contents of the universe, including this spiral galaxy, are moving away from one another. But will entropy go on increasing forever?

HEAT DEATH OF THE UNIVERSE?

When the idea of entropy was first developed, it was thought that the universe has a gloomy future ahead. Entropy would go on increasing until everything was at the same temperature: the stars would stop shining and their remains, the dead planets, and interstellar gas and dust, would all be at the same incredibly low temperature. Our inevitable fate trillions of years in the future would be a cold, dark, inert universe.

Today, the far-distant future of the universe seems more mysterious than ever before. For example, it has only recently been discovered that the expansion of the universe is accelerating, not slowing down as previously assumed. Scientists are no longer sure whether the idea of a universal "heat death" is true or not.

TAKING A CLOSER LOOK AT

THE CONCEPT OF REVERSIBILITY

An important concept in thermodynamics is that of reversible processes. Strictly speaking, no thermodynamic process is truly reversible, but scientists understand these processes best when they are as much like an ideal, thermodynamically reversible process as possible.

Reversible processes are smooth, gentle changes involving extremely small (theoretically zero) differences in pressure and temperature. An example is pushing a piston gently down in a cylinder of gas, slowly compressing it, and always with only slightly more pressure than the gas exerts. It takes only a very small change in pressure to reverse such a process. Slightly reducing the pressure on the piston means that the excess pressure of the gas will slowly push the piston out, a reverse of the original process.

Pushing down with a large pressure would not make a reversible process. It might, for example, send a shock wave through the gas. When the pressure is reduced the process would not reverse since no shock wave would appear in the gas and move up to hit the piston.

In normal use, compressing air in a bicycle pump is not a thermodynamically reversible process. That is because it is done with considerable power, generating heat and producing turbulence (movement) in the air inside.

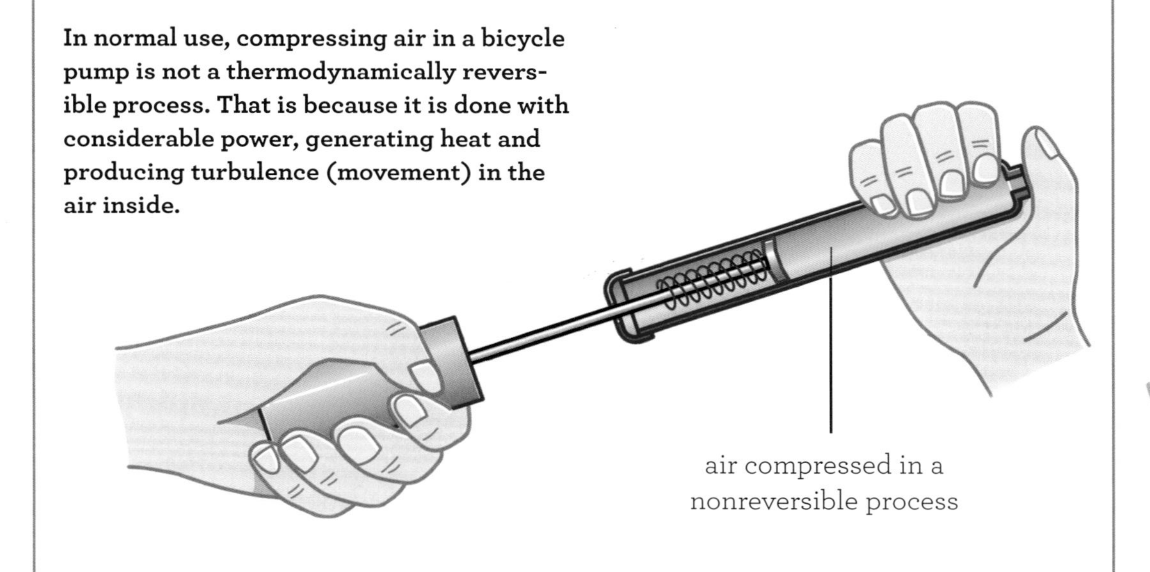

KEY DEFINITIONS

- **Absolute zero:** The lowest theoretically possible temperature, −459.67°F (−273.15°C).
- **Diffuse:** Spread out evenly, as heat and gases do.
- **Entropy:** A measure of the amount of disorder in any system.
- **Gibbs free energy:** A decrease in Gibbs free energy means that a reaction will happen spontaneously.
- **Kelvin scale:** Temperature scale that uses kelvins (K) as the unit of temperature, and where zero (0K) is absolute zero (−459.67°F; −273.15°C).
- **Microstate:** The state of a substance on the molecular scale—that is, the masses, speeds, and positions of all its molecules.

TEMPERATURE AND ENTROPY

When heat flows in a system, there is almost always an increase in the entropy of the system because heat is a disordering influence. The amount of entropy change is greater if the heat flow is greater. But the effect of heat in increasing disorder is also related to how orderly the system already is. The cooler something is, the more orderly it is, and the greater the amount of disorder that will be caused by a given amount of heat. So, for example, a given amount heat will cause a greater entropy increase in a mass of ice than in an equal mass of steam.

THE THIRD LAW

In addition to the first and second laws of thermodynamics, there is also a third law. This states that the entropy of matter is zero at the temperature absolute zero. Although absolute zero cannot be attained in practice, this law provides a definition of zero entropy, the least disordered state.

The average temperature in the universe is thought to be about 2.73K, or 2.73 degrees on the Kelvin scale (−454.76°F, −270.42°C). This is still a few degrees warmer than absolute zero.

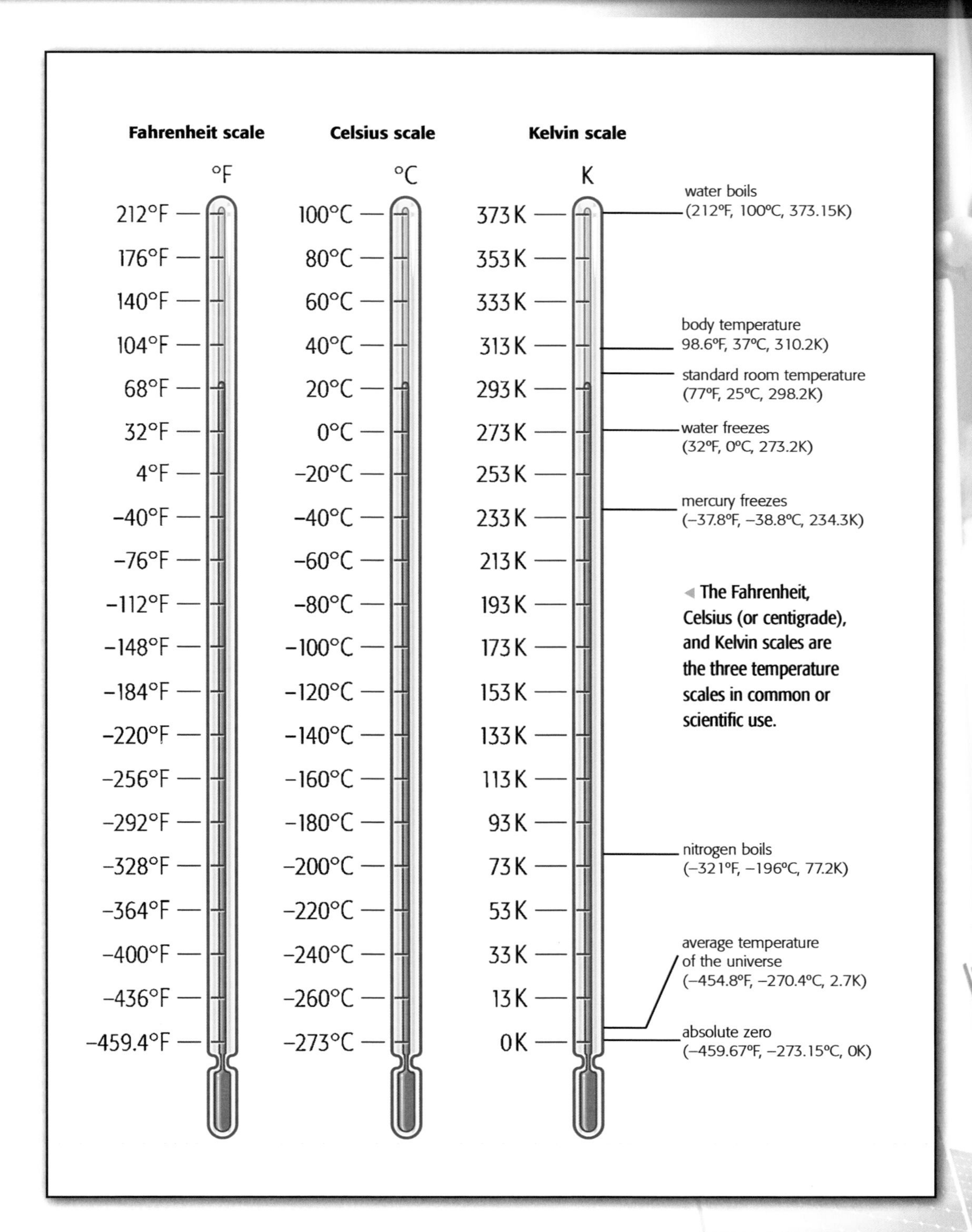

The Fahrenheit, Celsius (or centigrade), and Kelvin scales are the three temperature scales in common or scientific use.

WHAT'S ABSOLUTE ZERO?

Most people are familiar with the Fahrenheit or the Celsius scales, where temperatures are measured in degrees Fahrenheit (ºF) or degrees Celsius (ºC). Scientists, however, commonly use the Kelvin scale, with temperatures measured in kelvins (K). This scale uses intervals of the same size as the Celsius

When ice is immersed in water, it cools the water down until the mixture is at 32ºF (0ºC). It stays at this temperature until all the ice has melted. For this reason, melting ice provides a useful reference point for temperature scales.

A BRIEF PROFILE

LORD KELVIN

The 19th-century British scientist William Thomson did brilliant theoretical work in many areas of physics and was also a pioneer in many fields of engineering, including long-distance telegraphy. Thomson was honored with the title of Lord Kelvin for his scientific work. He devised the absolute (or Kelvin) scale of temperature, on which the unit of temperature is now called the kelvin, symbol K.

Pioneering scientist William Thomson (1824–1907) is better known today as Lord Kelvin.

(or centigrade) scale, where one degree is equal to one-hundredth of the difference between the temperature at which pure water boils and ice melts. (This is equal to nine-fifths of a degree on the Fahrenheit scale.) However, on the Kelvin scale, zero is at "absolute zero." On this scale, water freezes at 273.15K and boils at 373.15K. But what exactly is absolute zero?

As an object is cooled, the movements of its molecules slow down. Eventually a point would be reached at which the movements are as slow as they can possibly be. This point, which is at the extremely cold temperature of −459.67°F (−273.15°C), is called absolute zero.

Scientists first put forward the idea of absolute zero from studying gases. When a gas is cooled, its pressure decreases. This process comes to an end when eventually the pressure would fall to zero for a theoretical ideal gas. This lowest possible temperature is absolute zero.

Although it is not possible to reach exactly absolute zero in practice, scientists have been able to cool matter to just a tiny fraction of a degree above this temperature. At these very low temperatures, matter starts behaving very oddly indeed.

CHAPTER FOUR

REACTION SPEED

Some chemical reactions take place very fast, with dramatic effects. Others happen so slowly that chemists try to find ways to speed them up.

How fast reactions happen depends on the substances involved, and also on other factors that can be controlled, such as temperature and concentration.

In all chemical reactions, bonds between atoms are broken and new bonds are made. For example, in the burning of methane (CH_4) in oxygen (O_2), the bonds between the carbon, hydrogen, and oxygen atoms are all broken and new links are formed:

$$CH_4 + 2O_2 \rightarrow CO_2 + 2H_2O$$

When bonds change, electrons move between orbits in the outer layers of the atoms. This can happen only when the molecules or atoms come very close to each other, such as when they collide.

RATES OF REACTION

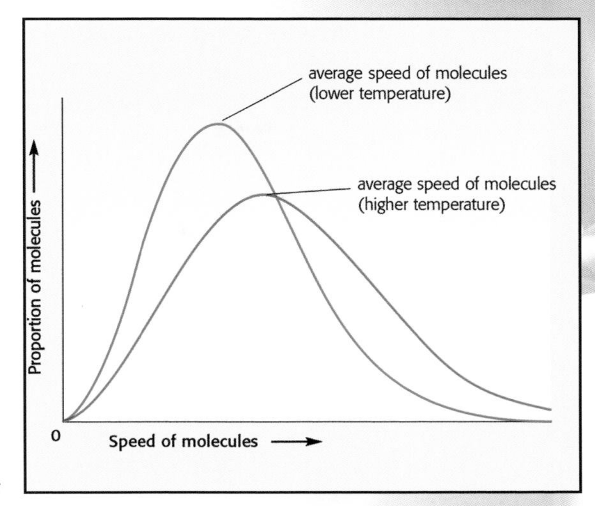

Molecules collide more often when they are more concentrated; that is, when there are more of them in a given volume. So at higher concentrations, the reactants have a greater chance of colliding and then reacting together. If a reaction involves two reactants and the concentration of one reactant is doubled, the reaction rate will be twice as fast. But if the concentration of both reactants are doubled, the rate will be four times as fast. Similarly, increasing the density of

As the temperature increases, the average speed of molecules increases, as does the proportion with higher speeds. The range of speeds also rises with temperature, so the graph is more spread out at the higher temperature (red line) than at the lower temperature (green line).

Gases are concentrated by compressing (squeezing) them. Underwater divers carry a concentrated supply of gas on their back for breathing.

gaseous reactants by compressing them usually helps a reaction go faster. At room temperature of 77 degrees Fahrenheit (77°F; 25°C) and normal air pressure, an oxygen molecule will travel on average 3 millionths of an inch (7.3 millionths of a centimeter) before colliding, and will have some 6.6 billion collisions per second. If the pressure is doubled, it will travel half as far between collisions on average, and will collide twice as often.

If the gas is at a higher temperature, its molecules move faster. Greater molecular speeds make a reaction go faster for two reasons. First, the faster molecules travel, the more frequently they will collide. However, this makes only a very small contribution to reaction rates. If the temperature of a gas (at atmospheric pressure) is raised from 77°F (25°C) to 95°F (35°C), there will be only about 2 percent more collisions per second.

Second and more important is the fact that, when molecules collide at higher speeds, they are more likely to interact.

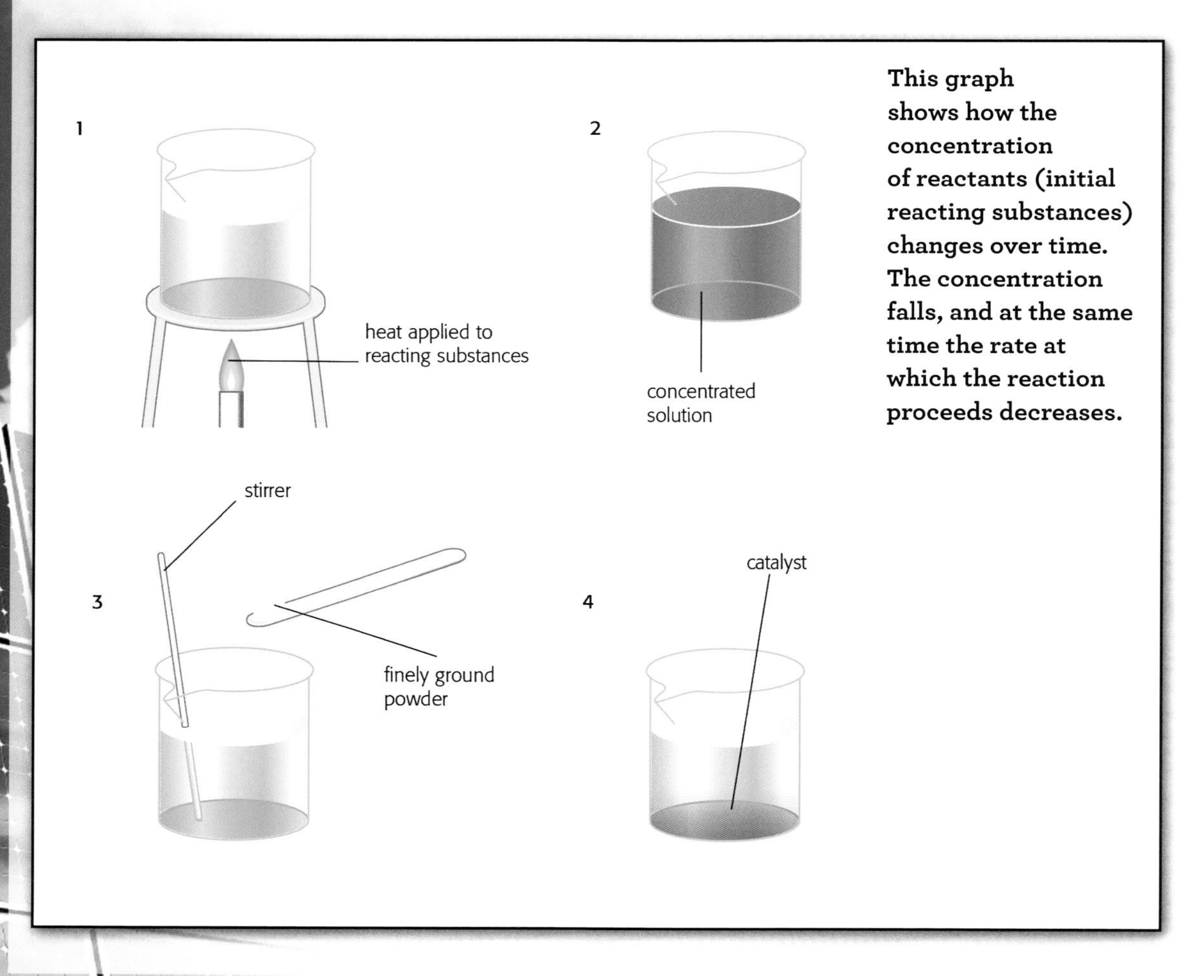

This graph shows how the concentration of reactants (initial reacting substances) changes over time. The concentration falls, and at the same time the rate at which the reaction proceeds decreases.

The molecules will have more energy to break and make bonds, rather than just bouncing off each other. However, at all temperatures there is a spread of molecular speeds, from very slow to very fast. There are always a few molecules traveling fast enough to react if they collide, even at lower temperatures.

Another factor influencing reaction rates is how easy it is for the reactants to mix. If all the reactants are liquids or gases, they will mix easily. But if one is a solid, it may need to be ground up into small pieces to increase the area of its surface exposed to the other reactants. A large lump of a solid substance will react more slowly than the same substance in powdered form. For this reason, chemists often use powdered chemicals, rather than lumps or large crystals.

Finally, adding a catalyst can speed up a reaction. Catalysts are substances that change the rate of a chemical reaction but are not used up in the process.

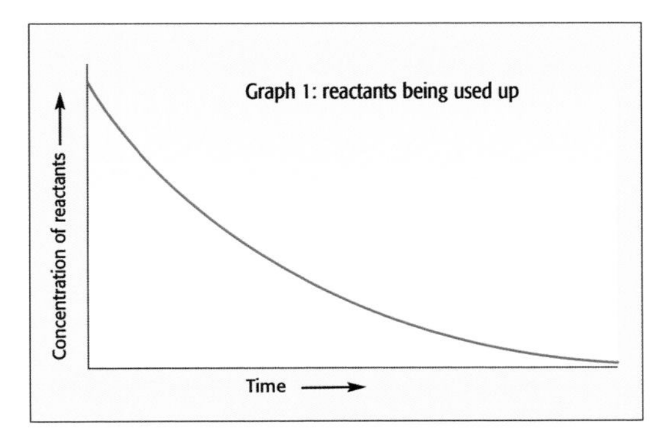

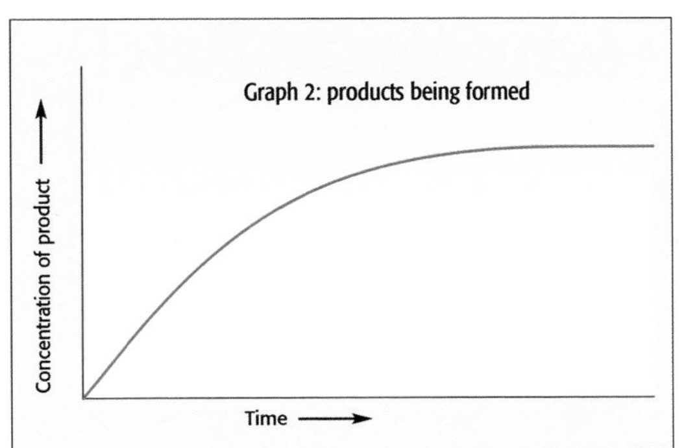

This graph shows how the concentration of products increases during the course of a reaction. The rate at which the reaction proceeds falls as the reactants are used up.

LOOKING AT REACTION CURVES

To measure how fast a reaction is taking place, we need to measure either how fast one or more of the products is being formed, or the rate at which a reactant is being used up. This means finding out how much of a product is present, or how much of a reactant is left, at different times. That is done by taking measurements at set intervals of time. For example, the compound hydrogen peroxide, H_2O_2, breaks down to form oxygen and water. (This reaction happens very slowly at room temperature, but it can be speeded up by raising the temperature or adding a catalyst.)

If we plot a graph of the measured concentrations of hydrogen peroxide against the time elapsed, this will make a curve that shows the rates of reaction as the breakdown process occurs.

On Graph 1, the concentration of hydrogen peroxide is highest at the start of the reaction. It gradually falls over time

COOL TOOLS AND TECHNIQUES

HOW TO MEASURE REACTION RATES

Scientists measure reaction rates by finding a characteristic of a product or reactant that can be measured, so they can monitor the concentration of that substance during the reaction. For example, if one of the products has a strong color, the rate of color change can be measured using an instrument called a colorimeter.

If a gas is one of the reaction products, this can be collected in a container and its volume measured at intervals. Some pieces of gas-collecting apparatus (such as gas syringes and eudiometers) allow the volume of gas inside to simply be read off a scale.

as the hydrogen peroxide breaks down. The slope of the curve is steepest at the beginning of the reaction, showing that the breakdown reaction is fastest when hydrogen peroxide levels are highest.

If we look at products being formed in a reaction, rather than reactants disappearing, the reaction curve will be a different shape. For example, imagine that two reactants react together to form two products. There, the reaction is also fastest at the beginning, where the curve is steepest. On this curve the concentration is rising because the curve shows the product being formed during the reaction rather than the reactant disappearing. As the reaction carries on, the reactants are used up and the rate slows

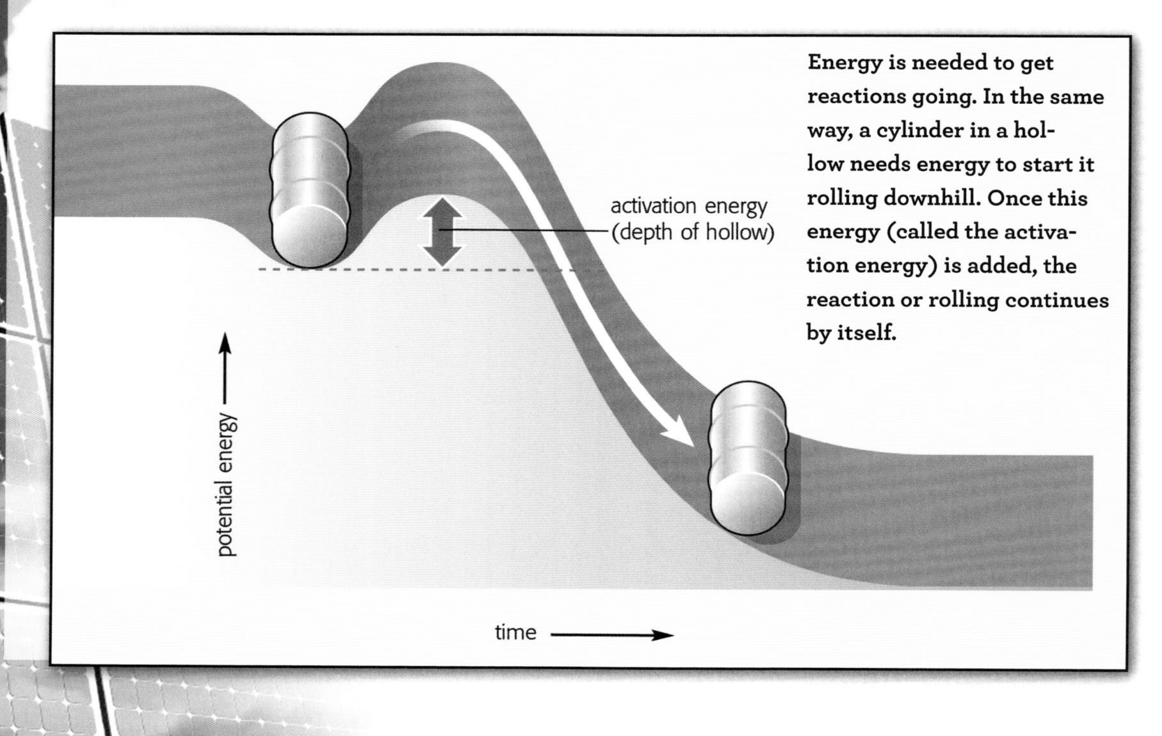

Energy is needed to get reactions going. In the same way, a cylinder in a hollow needs energy to start it rolling downhill. Once this energy (called the activation energy) is added, the reaction or rolling continues by itself.

down, so the curve levels off. Eventually no more products are formed, so the reaction curve becomes a straight, flat line.

UNDERSTANDING ACTIVATION ENERGY

When two molecules collide, they sometimes stick together to form a temporary molecule. These temporary, intermediate molecules, called activated complexes, often have high energy and cannot exist for long. Instead, they break down into product molecules or turn back into the reactant molecules.

This process is very important in reactions. Imagine that two molecules collide to form an activated complex. After a short time this breaks down to form the product molecules. To form the activated complex, the reactants need to take in a certain amount of energy. The difference between this energy and the average energy of the reactant molecules is called the activation energy. The larger this is, the more slowly the activated complex will be formed.

In this case, the reaction will not happen spontaneously, even if it is exothermic (that is, the products have less energy than the reactants). The molecules need to be supplied with the activation energy before the reaction can start. This is like an oil drum lying in a hollow at the top of a hill: it needs to be pushed out of the hollow before it can roll downhill. The energy that it is given to lift it to the top of the bump is like the activation energy needed to form the activated complex.

If a reaction as a whole is exothermic, energy is given out between the

Ammonia is the product of an exothermic reaction between nitrogen and hydrogen. It is produced and used in vast amounts.

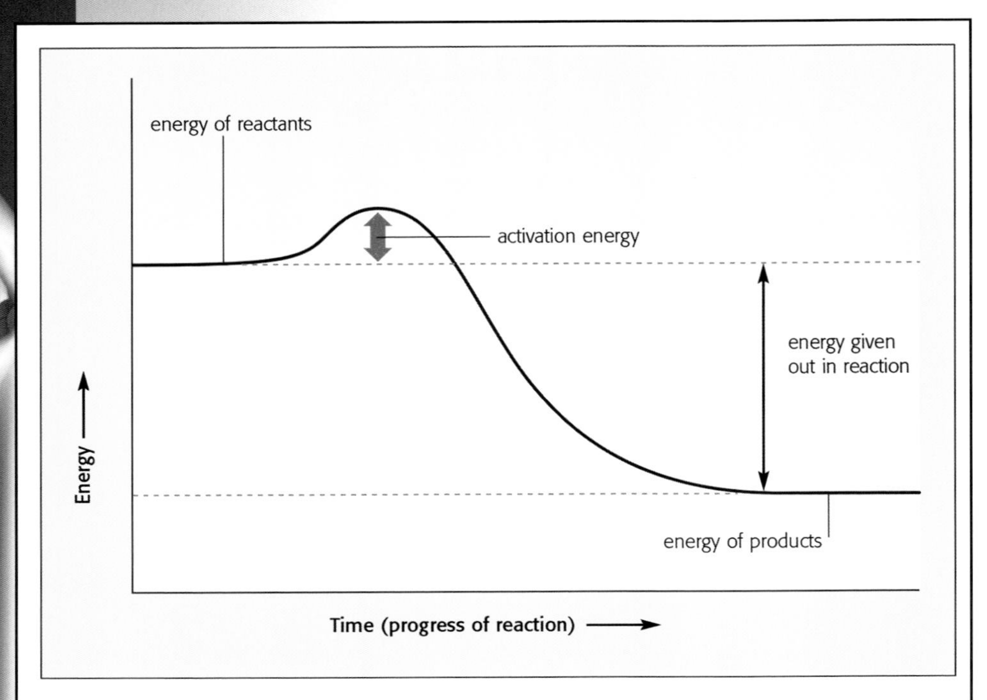

This graph shows the energy changes in an exothermic reaction. The reactants first form the activation complex. This compound has a higher energy than the reactants, so energy must be taken in. Energy is then given out when the activation complex breaks down, forming the products. The overall reaction is exothermic because the products have less energy than the reactants.

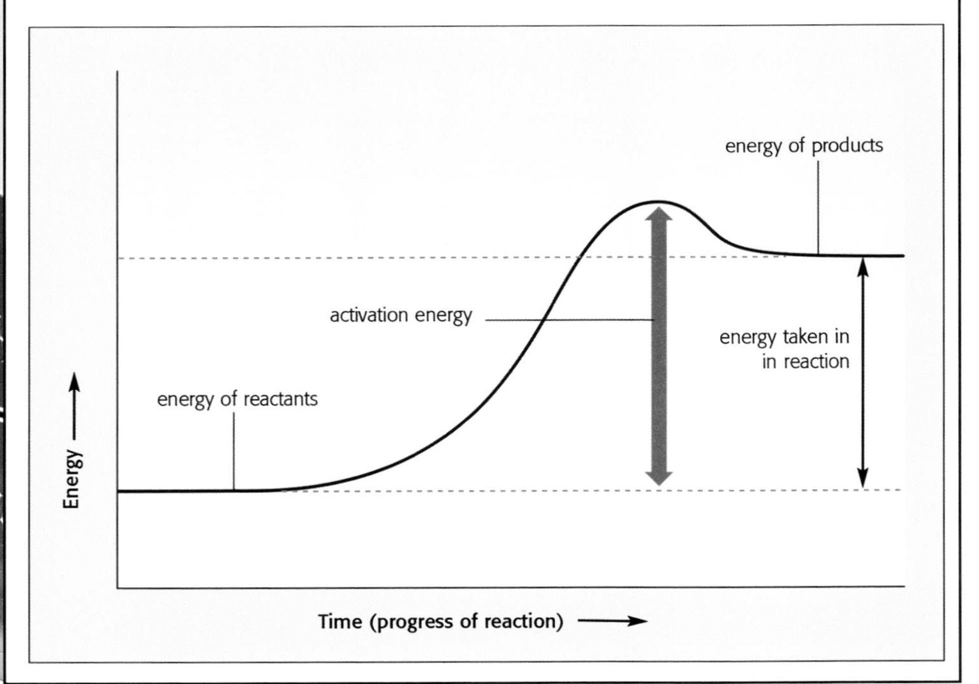

This graph shows an endothermic reaction. Here, energy is again needed to form the activation complex. When this breaks down, less energy is given out than was taken in because the final energy of the products is greater than that of the reactants.

TAKING A CLOSER LOOK AT

EVAPORATION AND FREEZING EQUILIBRIUM

Scientists measure reaction rates by finding a characteristic of a product or reactant that can be measured, so they can monitor the concentration of that substance during the reaction. For example, if one of the products has a strong color, the rate of color change can be measured using an instrument called a colorimeter.

If a gas is one of the reaction products, this can be collected in a container and its volume measured at intervals. Some pieces of gas-collecting apparatus (such as gas syringes and eudiometers) allow the volume of gas inside to be simply read off a scale.

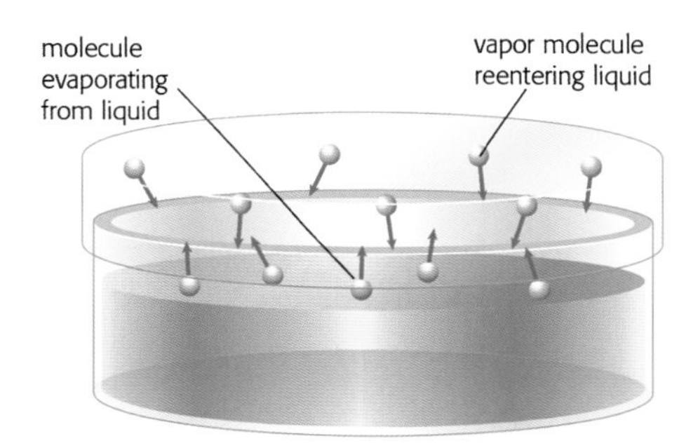

At equilibrium, as many molecules are moving back into the liquid as are moving out of the liquid into the vapor.

The element bromine is a liquid at room temperature. In a closed container, it forms an equilibrium with its brown-yellow vapor.

initial and the final stages. So, when the activated complex breaks down in an exothermic reaction, it releases more energy than it absorbed when it was formed. This is why the reaction as a whole gives out energy. However, if the chemical reaction is endothermic (absorbing energy), the activated complex releases less energy when it breaks down than was used in forming it. Many catalysts work by lowering the activation energy of a reaction, so the reaction can get going more easily and the products are formed faster.

UNDERSTANDING EQUILIBRIUM

If the conditions are right, many reactions go to completion. That is, they continue until the reactants are almost completely used up. However, other reactions do not continue so far. They reach a point where no more products seem to be formed, even though there are still reactants present.

In fact, what is happening is that the products are still being formed—but these are reacting together to produce the reactants again, at the same rate. When the reactants and products are being formed at exactly the same rate, the amount of each will stay the same.

At this point chemists say the reaction has reached equilibrium (balance).

An extremely important industrial reaction that reaches equilibrium is that of nitrogen gas (N2) with hydrogen gas (H2) to form ammonia (NH3):

$$N_2 + 3H_2 \rightarrow 2NH_3$$

This is called the forward reaction. At the beginning only hydrogen and nitrogen are present, but soon ammonia molecules are formed. As the ammonia molecules build up, they collide with each other more frequently and some break down into nitrogen and hydrogen:

$$2NH_3 \rightarrow N_2 + 3H_2$$

This is called the reverse reaction. It will go faster as the amount of ammonia increases, and eventually ammonia molecules will be breaking down as fast as new ones are made. At equilibrium there will be a mixture of nitrogen, hydrogen, and ammonia molecules.

To show that the reactions in both directions are occurring at the same time, chemists write this reaction using a double-headed arrow:

$$N_2 + 3H_2 \leftrightarrow 2NH_3$$

REVERSING REACTIONS

All chemical reactions can be reversed to some extent. Every reaction that we normally regard as having gone to completion actually contains a small amount of the initial reactants. But that amount can be incredibly small. The combining of hydrogen and oxygen to form water is regarded as irreversible because the spontaneous reverse breakdown of water is extremely slow. This in turn means that, once water has been formed from hydrogen and oxygen, hardly any of it will spontaneously break down again. The amount that will do so is extraordinarily small: in a volume of water equal to the entire Atlantic Ocean, only about five molecules will spontaneously break down into hydrogen and oxygen!

The ocean is a store of countless molecules of water, made up of hydrogen and oxygen atoms.

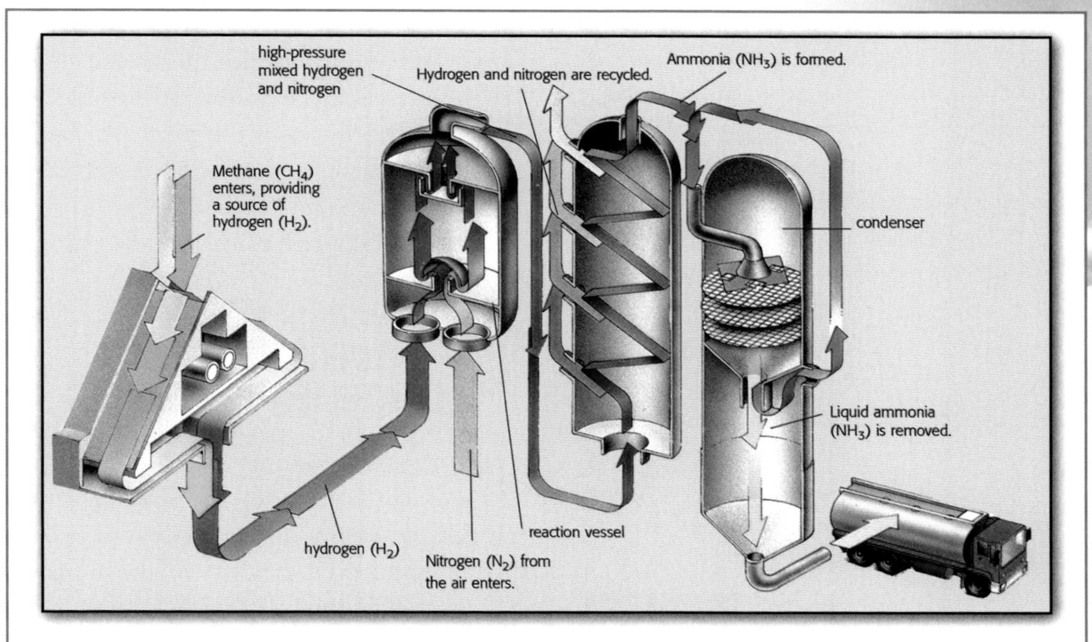

TAKING A LOOK AT
THE HABER-BOSCH PROCESS

Gaseous ammonia (NH_3) is a valuable industrial product. It is used in fertilizers, explosives, cleaning products, making batteries, and many other processes and products. Ammonia is made on a huge scale from nitrogen obtained from the air, plus hydrogen (mostly produced from naturally occurring methane). Before ammonia was produced by the Haber-Bosch process, the main industrial sources of nitrogen compounds were minerals that had to be mined and transported thousands of miles. By 1913, the research of two German chemists had made the industrial production of ammonia possible. Fritz Haber (1868–1934) studied the reaction in the laboratory, and Carl Bosch (1874–1940) of the company BASF turned it into an industrial process.

Haber worked on the reversible reaction:

$$N_2 + 3H_2 \rightarrow 2NH_3$$

The problem was that, as more ammonia was produced, it increasingly turned back into nitrogen and hydrogen. Bosch made this reaction go to completion by removing the ammonia as it was produced, through cooling and liquefying it. The unreacted nitrogen and hydrogen were recycled, so that eventually they were all used up.

THE EQUILIBRIUM POINT

If the temperature or pressure of a two-way reaction changes, the equilibrium amounts of the initial reactants and the final products will also change. Exactly how the equilibrium point is affected is complex, and it depends on the details of the particular reaction concerned. For example, in the production of ammonia, increasing the temperature causes less ammonia to form. By contrast, when carbon burns to form carbon monoxide, the equilibrium shifts toward increased production of carbon monoxide when the temperature increases.

Adding a catalyst will not alter the position of any equilibrium—that is, the amount of reactants and products. Catalysts speed up the rate of a reversible reaction in both directions. That is why they do not change the concentrations at equilibrium—although equilibrium can be reached sooner with a catalyst.

In any reaction, if the amount of products at equilibrium increases, chemists say that the equilibrium has shifted to the right (because the products are on the right of the equation). If a change in conditions makes more reactants form, they say that the equilibrium has moved to the left.

UNDERSTANDING LE CHÂTELIER'S PRINCIPLE

The French chemist Henri Louis Le Châtelier (1850–1936) formulated a rule that is very helpful in understanding how a chemical equilibrium will be affected by any changes. This rule states that, when a chemical system in equilibrium is disturbed, it changes in a way that tends to cancel out the disturbance.

For example, increasing the pressure makes gases dissolve or combine with other substances. So, the number of gas molecules is reduced. This tends to reduce the pressure, reducing the effect of the change in pressure. Similarly, increasing the temperature makes endothermic

Farmers still use natural sources of ammonia, such as manure, alongside industrial fertilizers containing ammonia. Manure spreading often takes place in winter, before the crops are planted.

KEY DEFINITIONS

- **Activated complex:** A short-lived molecule formed during a reaction that quickly breaks up into either the original reactant molecules or into product molecules.
- **Activation energy:** The difference between the average energy of reactant molecules at a given temperature and the energy they need to react.
- **Catalyst:** A substance that speeds up a chemical reaction but is not used up in the process.
- **Equilibrium:** In chemistry, the state of a chemical process in which the products are being formed as fast as they are breaking down, so the amounts of reactant and product are not changing.
- **Reaction rate:** The rate at which the concentrations of reactants and products change during the reaction.
- **Reversible reaction:** A reaction that can go both forward and backward. In other words, as well as the reactants forming the products (as in all reactions), the products react together to re-form the reactants in significant amounts.

(heat-absorbing) reactions go faster, which has the effect of reducing the temperature.

In the formation of ammonia from nitrogen and hydrogen, heat is produced (because the forward reaction is exothermic), and the volume decreases (because

Henri Louis Le Châtelier, whose principle explained how changing conditions affect chemical reactions. This principle has proved hugely important in industry, as well as in the laboratory.

two molecules of ammonia are produced from four molecules of nitrogen and hydrogen). Here, Le Châtelier's principle tells us that raising the pressure will shift the equilibrium toward the production of ammonia.

This is because ammonia takes up less volume than the reactants (nitrogen and hydrogen), so forming ammonia reduces the pressure. But raising the temperature will shift the equilibrium in the opposite direction—toward more nitrogen and hydrogen and less ammonia—so that less heat is given out and the temperature is lowered.

CHAPTER FIVE

CONTROLLING REACTIONS WITH CATALYSTS

Our lives depend on how fast chemical reactions proceed. Within our own bodies, countless different chemical reactions are going on at every moment. Most of these would happen far too slowly by themselves. In industry, many reactions are speeded up so that they produce a worthwhile amount of product within a reasonable time. Industrial chemists find ways to make reactions happen fast and safely.

A very useful way to speed up chemical reactions is to add a catalyst. Although a catalyst can help a reaction proceed faster, at the end of the reaction none of the catalyst will have been used up.

Petroleum is processed industrially in huge refineries. The process includes "cracking," in which large molecules are broken down, using catalysts, to form smaller ones.

At this brewery, the beer's alcoholic content is produced by yeasts, which ferment the natural sugars in the mix.

Yeast makes bread rise by catalyzing the production of carbon dioxide gas.

LOOKING AT CATALYSTS IN ACTION

People have been using catalysts in traditional foodmaking for thousands of years. In making beer, wine, and other alcoholic drinks, natural sugars such as glucose are broken down into carbon dioxide and ethanol (alcohol). The sugars are broken down by single-celled fungi called yeasts, which produce natural catalysts that promote these reactions. Yeast is also used in baking bread; the bubbles of carbon dioxide that yeast releases within the dough cause the bread to rise (swell up). Catalysts are just as important in the modern industrial processes used to produce much of the food we eat today.

In the petrochemical industry, a vast range of products are made from petroleum (oil extracted from the ground) using catalysts. The products include gasoline, lubrication oil, plastics, and gas for heating and lighting. All these materials are made from gases and liquids produced from petroleum. This mixture of compounds has molecules of different size. First, petroleum is heated so that the compounds with the smallest molecules can be driven off as gases; these are

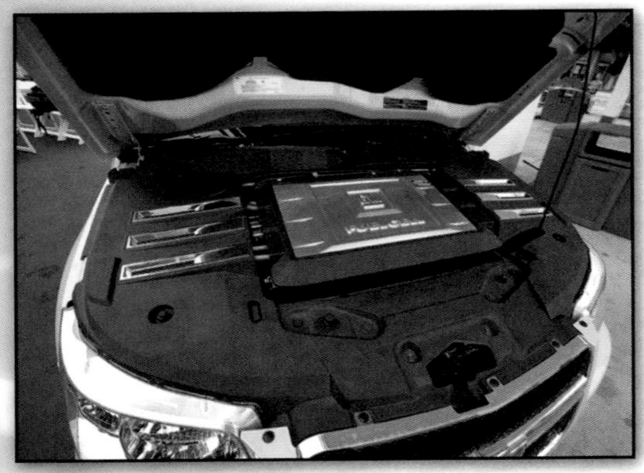

Fuel cells, already used in some cars, use platinum catalysts to produce energy from hydrogen and oxygen.

collected and separated from the remaining mixture. Then, the larger molecules in the mixture are chemically broken to form smaller, more useful molecules; this process is called cracking.

Other molecules may need to be joined together, or reshaped, to get the best blend of compounds. In these molecule-forming processes, chemical engineers use carefully chosen catalysts to control the types and amounts of products.

Catalysts are just as important in many other industrial processes.

In the Haber-Bosch process for making ammonia (see p. 51), finely divided iron mixed with oxides of potassium, calcium, and aluminum is used as the catalyst. Vanadium oxide (V_2O_5) is the catalyst used in the contact process for making sulfuric acid. Margarine is made by adding hydrogen to oils and fats; the metal nickel is used as a catalyst for this reaction.

Bars of high-purity platinum, an important catalyst metal.

UNDERSTANDING HOW CATALYSTS WORK

Some catalysts are gases or liquids that are mixed uniformly, or homogeneously, with reactants that are also gases or liquids. These are called homogeneous catalysts. They enable intermediate molecules called activated complexes (see p. 49) to form with a lower energy than the intermediate molecules in the unaided reaction. Because they have a lower energy, these intermediate complexes form more rapidly, and so do the final products.

Other catalysts are heterogeneous, which means that the catalyst is in one physical form—usually solid—and the reactants are in another (such as gases or liquids). The precious metals platinum and rhodium are used as heterogeneous catalysts in devices called catalytic

TAKING A CLOSER LOOK AT

HOW CATALYSTS AND INHIBITORS WORK

Catalysts are substances that increase the rate of a chemical reaction while not being altered themselves. Some compounds can slow catalyzed reactions by interfering with how catalysts work. Other compounds can slow reactions by tying up reactants so that they are not available. These compounds are called inhibitors. Inhibitors are used in home water-heating systems and in automobile cooling systems to slow down rusting.

Others are used in food additives to slow down the rate at which the food grows mold or goes bad.

Many fruits and vegetables contain antioxidants, which are inhibitors that slow down oxidation reactions in the body.

converters. These devices are used in automobiles to help the fuel burn completely and to break down poisonous gases in the exhaust fumes.

For example, nitric oxide (NO) is one of the gases that can cause pollution from car exhaust fumes. When nitric oxide molecules meet the metal in a catalytic converter, they are "adsorbed" onto it; that is, they become attached to its surface. On the metal surface, the nitric oxide molecules break down into individual atoms of nitrogen and oxygen. This may take place at only a small number of places, called active sites, on the surface of the catalytic converter. The nitrogen and oxygen atoms, being close to each other, can then combine to form nitrogen (N_2) and oxygen (O_2) molecules:

$$2NO \rightarrow N_2 + O_2$$

Another process that takes place in catalytic converters combines oxygen and poisonous carbon monoxide (CO) into carbon dioxide (CO_2), which is safe:

This gauze, made from platinum and rhodium, is used in nitric acid production. The reactions take place on the gauze's surface.

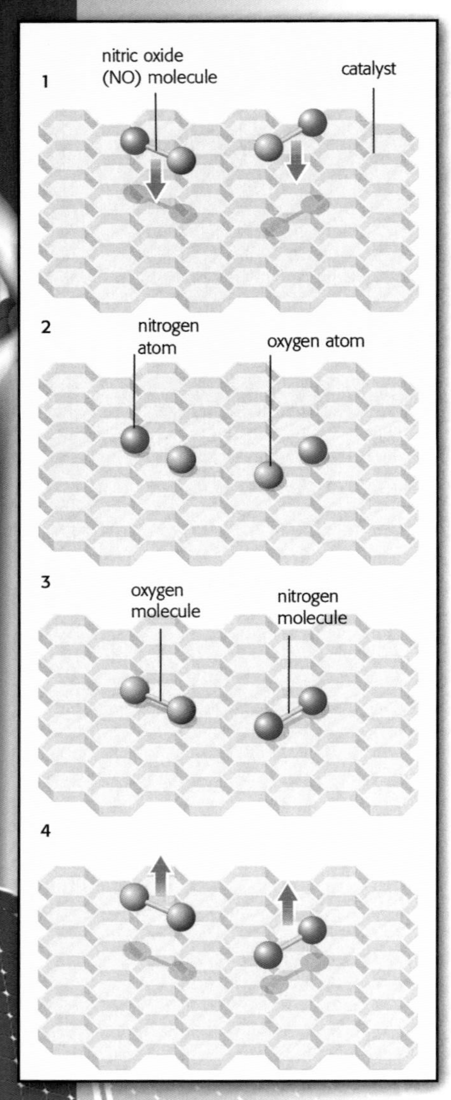

(1) In a catalytic converter, nitric oxide (NO) molecules arrive at the catalyst and attach to its surface.

(2) The nitric oxide molecules separate into atoms of nitrogen and oxygen.

(3) Molecules of nitrogen (N_2) and oxygen (O_2) gas are formed.

(4) The new gas molecules detach and move away.

$$2CO + O_2 \rightarrow 2CO_2$$

The catalytic converter also helps any traces of fuel present in the exhaust gases burn completely. For example, octane (C_8H_{18}) can be burned to form carbon dioxide and water:

$$2C_8H_{18} + 25O_2 \rightarrow 16CO_2 + 18H_2O$$

Notice that this involves the reaction of two octane molecules with no fewer than 25 oxygen molecules. On the surface of the catalytic converter, the reaction goes through a long series of steps. At each step, oxygen molecules come into contact with partially burned fuel molecules formed in the previous step, allowing further burning to take place. This means that the fuel is burned more completely, reducing pollution.

WHAT ARE ENZYMES?

The most extraordinary examples of catalysts are those called enzymes, which are found in living things. Enzymes are protein molecules, and like other proteins they are made up of long, folded chains of repeated chemical units called amino acids.

Enzymes make it possible for the body to break down food molecules,

Toxic fumes from cars are caused by partially burned fuel.

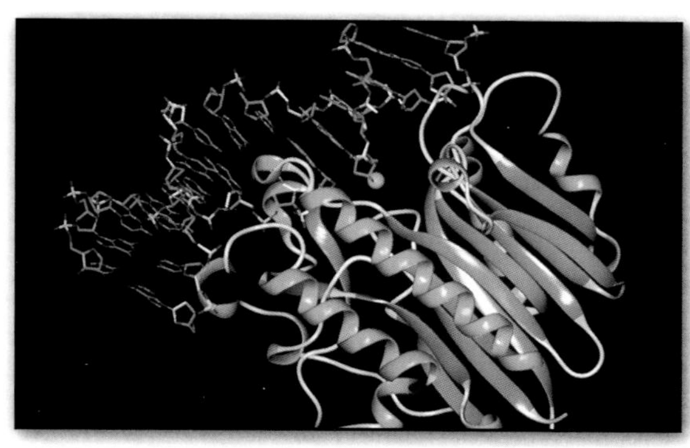

This enzyme molecule catalyzes the repair of DNA (deoxyribonucleic acid) molecules, which carry genetic information. DNA is constantly being damaged and repaired.

destroy toxic molecules, rebuild molecules needed elsewhere in the body, and do many other things. Their actions are incredibly precise: each enzyme targets particular molecules so that only a specific reaction is affected.

For example, amylase is a digestive enzyme present in human saliva. It breaks down starch to form maltose (a sugar consisting of two glucose molecules joined together). Similarly, the enzyme catalase, found in the blood, catalyzes the breakdown of harmful hydrogen peroxide (H_2O_2), which can build up in the body as a result of natural processes, to form oxygen and water. Hydrogen peroxide is also sometimes used on skin wounds as a disinfectant: the enzyme's action makes it bubble vigorously.

UNDERSTANDING SUGARS, ENZYMES, AND MILK

Our bodies need to obtain the sugar glucose from our food. Food often contains complex sugars, in which the molecules consist of simple sugars (such as glucose) joined together. For example, human and cow's milk contains lactose, in which glucose is combined with another simple sugar, galactose. The enzyme lactase in the human body breaks down lactose into glucose and galactose molecules.

Some people do not make enough lactase to break down lactose. If they drink milk, they feel ill. They can drink soy milk instead, which is made from the soy plant. In fact, lactose intolerance is surprisingly "normal": the people of many ethnic groups in the world can only digest milk in childhood and lose this ability when they become adults.

HOW DO ENZYMES WORK?

Enzymes work by a "lock-and-key" mechanism. Each enzyme molecule has a particular place on it that acts as the "lock." Certain reactant molecules are like keys: these molecules (and no others)

Milk provides an important food source. However, in some of the world's populations, adults cannot digest milk.

LET'S TRY THIS

GLUCOSE TESTING IN MILK

Materials: a beaker or glass; some plastic cups; ordinary cow's milk; soy milk (or any other lactose-free "milk" such as rice milk); and glucose powder or tablets. You will also need some lactase drops and glucose test strips (preferably for testing for glucose in urine rather than blood), which are available in most drugstores.

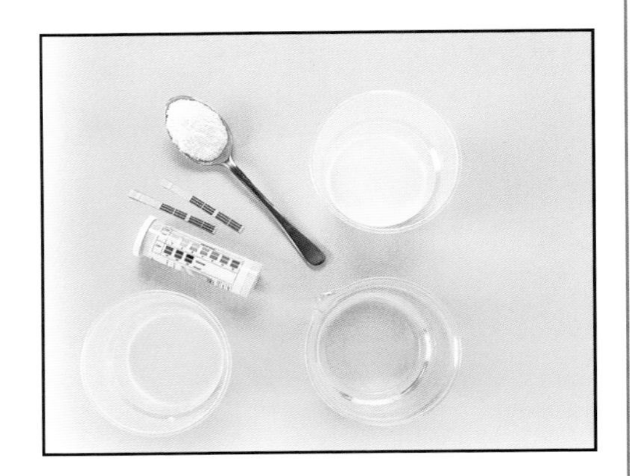

First, pour half a cup of water into the beaker or glass. Add a tablespoon of glucose and stir well, then dip a glucose test strip into the solution. Compare its color with the key on the package to find out the amount of glucose in the solution (and write down the result).

Next, put the soy milk into a cup, test with a fresh test strip and write down the result. Then put some cow's milk into a cup and test that with another strip, noting the result. Finally, add a few lactase drops to the ordinary milk, stir it up, test it again with another strip, and note the result.

Looking at your results, which liquid had the most glucose and which the least? How did the glucose level change when you added the lactase drops to the milk?

Comparing the colors on the test strips with those on the key reveals the concentration of glucose in each of the beakers.

You should have found that there was almost no glucose in the ordinary milk (and much less than in the glucose and water mixture). This is because ordinary milk contains little glucose; its sugar is in the form of lactose. Adding lactase breaks down lactose into glucose and galactose, so there should be more glucose in the milk and lactase mixture than in the ordinary milk. There should be no glucose in the soy milk, which contains only other sweeteners.

TAKING A CLOSER LOOK AT

THERMOMETERS THAT ARE ALIVE

Enzymes control many behaviors in living creatures. One example is the rate at which crickets chirp. Male crickets make their "song" by rubbing their wings or legs together. When the air temperature changes, this alters the rate of enzyme-controlled reactions in the crickets' nervous systems, which changes the rate of chirping. A cricket typically chirps 60 times a minute at 55 degrees Fahrenheit (55°F; 13°C). This rate rises as temperature rises.

fit into the "lock" region of the enzyme, which is the active site. The "key" molecules are called substrates.

When an enzyme catalyzes a reaction, the enzyme molecule is in contact with all kinds of reactant molecules. Only the substrate molecules that encounter the active site are adsorbed (stick) there. In a breakdown reaction, the enzyme structure helps the bonds in the adsorbed substrate molecules break.

The molecules can then split apart, and finally the parts separate from the enzyme molecule.

In a reaction where two substrates combine, a molecule of one of the substrates will be first to reach the enzyme's active site and stick there. Later, a molecule of the second substrate will encounter the enzyme at the right place and be adsorbed. Having been brought together, the two substrate molecules react with each other, forming the product molecule. That then separates from the enzyme.

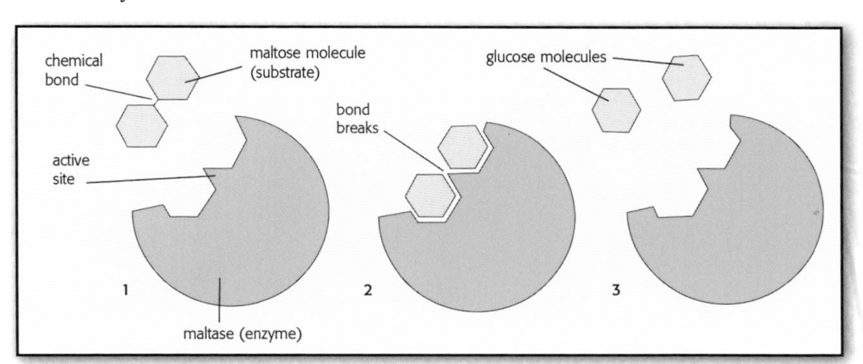

The enzyme maltase breaks down the complex sugar maltose to form glucose. A maltose molecule approaches the enzyme (1) and attaches to the active site (2). After reaction, the glucose molecules are released (3).

USING ENZYMES IN TECHNOLOGY

Enzymes are essential to every living organism, and scientists and engineers are increasingly finding uses for enzymes in industry and advanced technology. For example, enzymes are now used in making laundry powders, food, cosmetics, and medicines. One advantage of using enzymes to catalyze industrial reactions is that they can make reactions happen under mild conditions, rather than at the high temperatures and pressures often used in industry. Another advantage is that enzymes can be much cheaper than the hugely expensive precious metals that are often used as catalysts.

An exciting possibility is that enzymes may become a vital part of nanotechnology, in which machines are built on the molecular scale. Already

TAKING A CLOSER LOOK AT

JUST THE RIGHT TEMPERATURE

The catalytic activity of an enzyme is reduced or destroyed if the temperature is too cold or too hot. Warm-blooded animals, such as humans, are those that can control their internal temperatures to keep them in the right range for enzyme activity. "Cold-blooded" animals, such as reptiles, are not really cold-blooded: they, too, have to keep the enzymes in their system working by keeping the inside of their body reasonably warm. They do this by generating heat through muscular activity, or by moving into sunshine or shade as needed.

This iguana basks in the sun on a warm rock. This helps keep its body temperature right for enzyme activity.

CHEMISTRY IN ACTION

GETTING CLEAN

"Biological" laundry powders contain enzymes that attack the proteins and fats in stains on clothes, often caused by spilled food. Three types of enzymes are commonly used: proteases (which digest proteins), amylases (which break down starch), and lipases (which break down fats). Together, these enzymes make up a very small proportion of the washing powder: less than 1 percent in total. The enzymes used are produced by bacteria. Enzymes make it possible to get clean washing using a low temperature (around 104°F; 40°C), saving on energy costs. But the enzymes themselves have to work in a much harsher environment than that inside living things: they need to withstand detergents, soaps, and chemical oxidants, all in temperatures that can be significantly above body temperature.

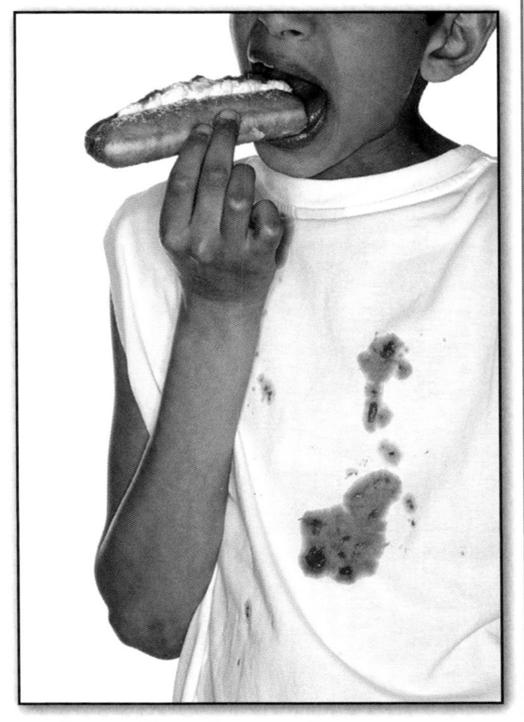

Biological laundry powders contain food-digesting enzymes that help remove messy stains on clothes.

"DNA computers," which use enzymes and can make calculations in a test tube, have been built. Some researchers think that, in the future, tiny enzyme-and-DNA computers could be embedded in our body to monitor health and release drugs to repair damaged or unhealthy tissue.

KEY DEFINITIONS

- **Active site:** The place on an enzyme or other catalyst where the reactants attach and the reaction occurs.
- **Adsorption:** The process of molecules becoming attached to a surface.
- **Enzyme:** A biological protein that acts as a catalyst.
- **Inhibitor:** A substance that slows down a chemical reaction without being used up by it; also called a negative catalyst.
- **Protein:** A substance consisting of large molecules made up of chains of chemical units called amino acids.
- **Substrate:** The particular molecule on which an enzyme acts.

CHAPTER SIX

BIOGRAPHY OF ISAAC NEWTON

"Nature and nature's laws lay hid in night: God said, Let Newton be! and All was light."
— Alexander Pope (1688–1744)
Epitaph written for Newton

Isaac Newton's beginnings gave little indication of how important he would become in the future. The son of a farmer in the village of Woolsthorpe in Lincolnshire, eastern England, Newton had a difficult early life. He was born prematurely on Christmas Day, 1642, about three months after his father's death, and was described as small enough to fit into a two-pint pot. When he was three years old, his mother remarried and left Isaac with his grandmother while she moved to her new husband's home in a nearby village. She did not return until her second husband's death eight years later. Newton appears to have experienced some bitterness about this. One of his notebook entries confesses that he had threatened "my father and mother Smith to burne them and the house over them."

A TALENT FOR INVENTION

As a youth, Newton revealed considerable skill in constructing working mechanical models. A windmill he built, for example, was operated by a mouse running on a treadmill. He also designed numerous sundials, developing in the process an ability to tell the time by the sun with impressive accuracy. At one time he became so distracted by his design for a model waterwheel that he allowed the sheep

It was supposedly while sitting in his garden at Woolsthorpe that Newton began to wonder whether the same force that makes apples fall from trees keeps the moon in orbit around Earth. The scene is recreated in this 19th-century painting.

Woolsthorpe Manor, Lincolnshire, Newton's family home. The house still stands today and is preserved as it would have been in Newton's time.`

KEY DATES

1642	Born December 25 at Woolsthorpe Manor, Lincolnshire
1645	Remarriage of mother Hannah to Reverend Barnabas Smith
1653	Mother returns to Woolsthorpe after death of second husband
1655	Attends King Edward VI Grammar School, Cambridge
1661	Admitted into Trinity College, Cambridge
1672	Publishes *A New Theory on Light and Colors*
1679	Death of mother
1687	*The Mathematical Principles of Natural Philosophy*, known as the *Principia*, describes Newton's theory of planetary motion
1693	Suffers nervous breakdown
1696	Appointed warden of Royal Mint in London; becomes master of the Mint in 1699
1703	Elected president of the Royal Society, London
1704	Major work on light and color, *Opticks*, published
1705	Knighted by Queen Anne to become Sir Isaac Newton
1727	Dies March 20; later buried in Westminster Abbey

he was supposed to be watching to stray into a neighbor's field.

As an only child, young Isaac inherited his father's farm; it was assumed that he would take over the running of it, as his mother wished him to do. However, the boy clearly had other interests and, in 1661, at the suggestion of her brother, Newton's mother agreed instead to send her son to Trinity College, Cambridge. Here, Newton developed a particular passion for mathematics and astronomy. As at other universities, the curriculum was based principally on Aristotle's ideas, but Newton began to study independently the works of more modern philosophers, such as the Englishman Francis Bacon (1561–1626) and the Frenchman René Descartes (1596–1650).

A PRODUCTIVE STREAK

By the summer of 1665 the Great Plague, which had broken out in London and eventually killed more than 75,000 people, had reached Cambridge. The university was

Newton's experiments revealed that as white light passes through a prism—a triangular block of pure glass—the beam divides into the colors of the spectrum. The blue end of the spectrum is refracted most strongly, and the red least; we now know that this happens because each color band has a unique range of wavelengths.

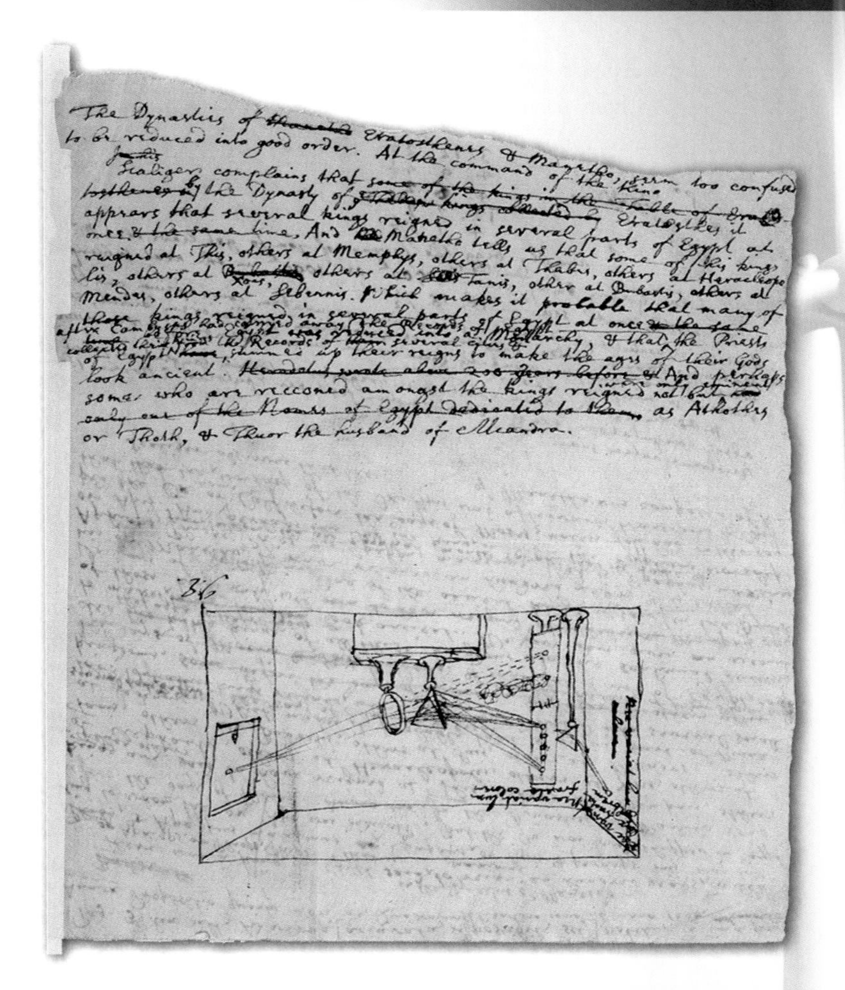
The Dynasties of Eratosthenes & Manetho, seem too confused
to be reduced into good order. At the command of the king
Scaliger complains that ... by the Dynasty of ... Eratosthenes it
appears that several kings reigned in several parts of Egypt at
once. And Manetho tells us that some of his kings
reigned at This, others at Memphis, others at Thebes, others at Heracleopo-
lis, others at Xois, others at Tanis, others at Bubastis, others at
Mendes, others at Sebennis. Which makes it probable that many of
those kings reigned in several parts of Egypt at once ... & that the Priests
of Egypt ... summed up their reigns to make the ages of their Gods
look ancient. And perhaps
some who are recconed amongst the kings reigned not ... as Athothes
or Thoth, & Thuor the husband of Alcandra.

A sketch by Newton illustrates how light is refracted by a prism. His work on colors and light was fully described in his book, *Opticks* (1704).

closed. Newton returned home to Woolsthorpe and spent the whole of the next year there. The year 1666 is sometimes described as Newton's year of wonder ("annus mirabilis"). He was, he said later, at the height of his inventive powers, and more in love with mathematics and philosophy than at any other time of his life. He first developed his theories of color and gravity at this time, as well as a range of new mathematical ideas. Yet he kept much of this work to himself, showing just a part of it to colleagues later. The rest was only published after his death.

By 1669, Newton had returned to Cambridge and become professor of mathematics there. His abilities began to be recognized when he designed a radical new telescope, which he presented to the Royal Society, later to become the world's leading scientific institution. This so impressed the Society that in 1672 it elected him a member.

PLAYING WITH LIGHT

In the same year Newton wrote a letter to the Royal Society setting out *A*

New Theory on Light and Colors. It met with considerable criticism, particularly from the distinguished physicist, Robert Hooke (1635– 1703). Newton was deeply affected by this and returned to working on his own; after this he was always slow to reveal his discoveries and reluctant to publish his findings.

Newton continued his experimental work with color, however. At this time, scientists did not believe that color is a basic property of light, but thought it was a modification that occurred when light passed through another substance such as water, cloud, or glass. Newton, however, became intrigued by his observations of what happened when light passes through prisms—triangular blocks of glass.

When light rays pass between substances of different densities, such as air and water, they change direction. This is called refraction. It is refraction that makes a stick half in and half out of water look as if it is bent or broken. Newton noted that a thread, colored half red and half blue, when pulled tight and viewed through a prism, seemed to split in half; the blue half appearing a little higher than the red. Newton wondered whether the rays that made blue appeared higher than those that made red because they were refracted more.

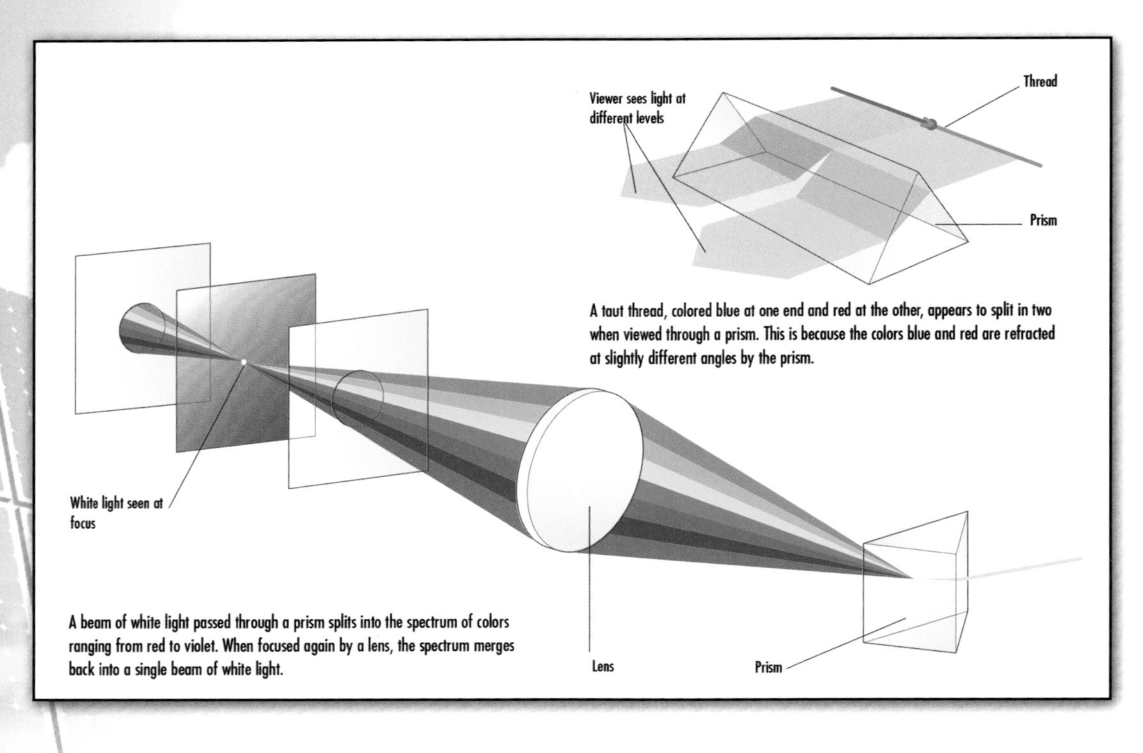

A taut thread, colored blue at one end and red at the other, appears to split in two when viewed through a prism. This is because the colors blue and red are refracted at slightly different angles by the prism.

A beam of white light passed through a prism splits into the spectrum of colors ranging from red to violet. When focused again by a lens, the spectrum merges back into a single beam of white light.

BREAKING DOWN LIGHT

Newton tested his idea with further experiments, and in 1672 made a series of groundbreaking observations. In a darkened room he passed a beam of light through a prism onto a screen. As the light traveled through the prism, it fanned out into the colors of the spectrum. These are the colors seen in a rainbow, ranging from red at one end to violet at the other. Newton demonstrated that if a beam from a single color from the spectrum—red, for example—was passed through a second prism, this light remained red: it could not be changed again.

When the second prism was replaced with a lens and the colors of the spectrum passed through this, they turned to white light again.

Newton drew a number of conclusions from these experiments. He decided that color must be a basic component, or part, of light, and that white light was the result of combining all the colors of the spectrum. He also found that every color in the spectrum has a different and unique degree of refraction. We now know that this is because light is made up of waves. Each color band has a unique range of wavelengths, and each color band is always refracted the same amount by a prism. Red light, for instance, has a longer wavelength than violet light, and so is refracted less. Newton, however, thought that light was made up of small particles that he called "corpuscles." These, he suggested, were ejected from the source of light like bullets from a gun.

Newton's views on the nature of light were challenged by Robert Hooke, who had also made a study of optics, and by the Dutch physicist Christiaan Huygens (1629–1693). They both favored the wave theory of light, but Newton firmly resisted their arguments. He pointed out that while we know that sound waves are able to bend because we can hear a bell ringing from the other side of a hill, light does not behave in the same way—the hill prevents us from seeing the church tower in which the bell is ringing. In fact, we now know that light sometimes behaves like particles and sometimes like waves.

As so often in his life, Newton was angered by criticism of his views. His reaction to Hooke's and Huygens's objections was extreme: he refused to publish anything more. Not until 1704, more than 30 years later, did the results of his work on light and color appear—with some additions and revisions—in his book *Opticks*. Meanwhile, he had turned to consideration of how things move.

THE THREE LAWS OF MOTION

Newton eventually succeeded in formulating three laws of motion, which he wrote up in 1687. As already noted,

This device is called a "Newton's Cradle," and demonstrates the conservation of energy as the spheres transfer energy to one another.

the first states that "a body continues to move with a constant velocity [speed and direction] or to...remain at rest unless acted upon by an external force." In other words, a moving object continues to move at a steady speed in a straight line, or if at rest continues to stay at rest, unless interrupted.

The second law states that when a force is applied to an object, it accelerates in a straight line in the direction of the force. The object's acceleration (the rate of change in its velocity) depends on its mass (the amount of matter it contains). The less the object's mass, the greater the acceleration (a light stone is easier to throw than a heavy one).

Newton's third law maintains that if an object exerts force on another object (the action), then that second object exerts an equal but opposite force back (the reaction). If you push or pull an object, it will push or pull you back to an equal extent. So when an astronaut pushes against a spaceship with a certain force, the spaceship pushes against the astronaut with an exactly equal force.

PHILOSOPHIÆ
NATURALIS
PRINCIPIA
MATHEMATICA.

Autore *J S. NEWTON*, *Trin. Coll. Cantab. Soc.* Matheſeos Profeſſore *Lucaſiano*, & Societatis Regalis Sodali.

IMPRIMATUR.
S. PEPYS, *Reg. Soc.* PRÆSES.
Julii 5. 1686.

LONDINI,
Juſſu *Societatis Regiæ* ac Typis *Joſephi Streater*. Proſtant Venales apud *Sam. Smith* ad inſignia Principis *Walliæ* in Cœmiterio D. *Pauli*, alioſq; nonnullos Bibliopolas. *Anno* MDCLXXXVII.

This is the title page to Newton's most famous work, *Principia*.

In the 20th century, it was discovered that Newton's laws of motion do not apply in the area of quantum mechanics (the physics of very smallscale or high-speed particles and atoms). But the laws of motion formulated by Newton more than 300 years ago hold good for all objects moving at normal speeds.

GRAVITATION

In outline, Newton's assertions were incredibly simple. First he stated that there was only one force at work—a force of attraction. This is what we call the force of gravitation, and Newton showed that it governs all motion, whether in space or on Earth. Every piece of matter in the universe attracts every other piece. The force of gravitation operates in a surprisingly neat way, depending only upon the masses of the bodies involved and the distance between them. All other factors can be ignored. It does not matter if one object is hot, the other cold, or if one is made of iron, the other of copper. As a further simplification, Newton demonstrated that the attractive force of a spherical object was concentrated at the center of the sphere.

In later life Newton reported that the basic theory had come to him while he was sitting in the orchard of his home at Woolsthorpe in 1666, his year of wonders. The famous story is that he watched an apple fall to the ground. He suddenly realized that the power of gravity that brought the apple down need not be limited to any particular distance from Earth but might extend much farther. "Why not as high as the moon?", he is supposed to have asked. Newton realized that just as Earth's gravitational pull attracts an apple, making it fall to the ground, so it attracts the moon and keeps it in orbit.

How does gravity cause the moon to orbit Earth? As Newton had shown by his first law of motion, if Earth were not there, then the moon would move through space in a straight line and at a steady speed. However, a balance is achieved between the moon's being pulled down toward Earth by Earth's gravitational pull and the moon's forward motion along its original path. It is this balancing process that deflects the moon sideways and keeps it in orbit around Earth.

As with Earth and the moon, so with the sun and the planets. Newton argued that the force between any two bodies (the sun and a planet) varies depending on their mass and the distance between them. He calculated that the attractive force of gravity increases with mass and weakens with distance, operating according to an "inverse square" law. So a planet twice as far away as another would be pulled with a quarter of the force, a planet four times as far away with one-sixteenth of the force. This inverse square law allowed Newton to prove mathematically what Kepler had observed, that the planets move in elliptical orbits. The French writer and enlightenment figure Voltaire (1694–1778) was a great admirer of Newton and helped to publicize his theories in France. He wrote: “Before Kepler all men were blind. Kepler had one eye, and Newton had two eyes,” meaning that Kepler saw half the truth about the universe, but Newton saw it whole.

WELL-DESERVED SUCCESS

In 1693, Newton appears to have suffered a serious breakdown. Some historians think his symptoms may be an indication of mercury poisoning. Mercury, a highly toxic metallic element that can cause dementia, was much used in alchemy, the precursor of modern chemistry. Newton—whose interests embraced all branches of science and philosophy—is known to have studied alchemy in great depth and to have conducted chemical experiments throughout his life. However, others point to stress in his private life as a more likely cause of his breakdown. He appeared to observers to be “in a frenzy,” accusing his friends of plotting against him and trying to entangle him with women. The illness, whatever its cause, seems to have passed within the year.

By now it was clear that Newton desired to see his career marked by appointment to a significant public position. In 1696, an influential friend was able to secure for Newton the post of warden at the Royal Mint, the body responsible for producing all the English coinage and medals. He remained at the Mint for the rest of his life, being promoted to master in 1699.

These were busy years at the Mint, as England was undergoing a great reform of its coinage. The old silver coins had a smooth edge, making it easy for people to clip small amounts of silver from them and still be left with usable coins. Counterfeiting (forging false coins) was also common. To avoid both these problems, it was decided to put new coins with milled or grooved edges into circulation. Newton oversaw the change from the old to the new coinage; he also took a leading role in the campaign against counterfeiters, visiting taverns and prison cells to take statements from prisoners and informers. Newton carried out his duties with ruthless efficiency: in 1697 alone, 19 counterfeiters were executed under his orders.

Newton served two terms as an English member of Parliament, and was elected president of the Royal Society in 1703. He was knighted in 1705 by Queen Anne for his public duties. In the last years of his life, he published further editions of his principal works, with some additions and alterations, but he did not undertake any major new scientific work.

“If I have seen further [than others],” Newton once said, “it is by standing on the shoulders of giants.”

On another occasion, he summed up his achievements in a striking image: “I do not know what I may appear to the world, but to myself I seem to have been only like a boy playing on the seashore, and diverting myself in now and then finding a smoother pebble or a prettier shell than ordinary, whilst the great ocean of truth lay all undiscovered before me....”

SCIENTIFIC BACKGROUND

POLITICAL AND CULTURAL BACKGROUND

Before 1660

Greek philosopher Aristotle (384–322 BC) argues that things on Earth move when pushed or pulled by a "mover"

German astronomer Johannes Kepler (1571–1630) publishes his first two laws of planetary motion in *The New Astronomy* (1609), revealing that planets move in elliptical paths

Italian mathematician, physicist, and astronomer Galileo Galilei (1564–1642) publishes his *Discourses upon Two New Sciences* (1638), in which he presents his laws of motion and friction, contradicting many of Aristotle's assertions

French philosopher René Descartes (1596–1650) sets out his view of the cosmos in *Principles of Philosophy* (1644)

1660

1662 The Royal Society in London is granted its royal charter

1665–66 Newton develops ideas on calculus, light and optics, and gravity

1665 The New Jersey colony is founded by English colonists, who make Elizabethtown their capital

1668 Newton invents the reflecting telescope

1670

1672 English physicist Robert Hooke (1635–1703) suggests a "wave" theory of light

1675 English astronomer John Flamsteed (1646–1719) is appointed astronomer-royal at the Greenwich Observatory in London, England by King Charles II (1630–1685)

1675 Work begins on St. Paul's Cathedral in London, England. It is designed by English architect, astronomer, and mathematician Christopher Wren (1632–1723)

1679 Hooke suggests that gravitational force might operate according to an "inverse-square" law of attraction, leading to a later dispute with Newton

1680

1682 English religious nonconformist William Penn (1644–1718), member of the Society of Friends (Quakers), is granted land in North America by King Charles II to establish a Quaker colony; the territory is called Pennsylvania

1683 William Penn signs the Great Treaty of Shackamaxon, by which the Delaware Native Americans grant him vast territories

1684 German mathematician Gottfried von Leibniz (1646–1716) invents calculus, a branch of mathematics, prompting a long-running argument with Newton

1687 Newton presents his three laws of planetary motion and his law of universal gravitation in his groundbreaking book, the *Principia*

1690

1690 Dutch physicist Christiaan Huygens (1629–1693) publishes his theory of gravity in *Discourse on the Cause of Gravity*

1690 Huygens publishes his *Treatise on Light*, which has been almost complete since 1678. In it he explains reflection and refraction, and puts forward his wave or pulse theory of light

1690 William III of England (1650–1702) completes the Protestant conquest of Ireland when he defeats the Catholic ex-king of England, James II (1633–1701), at the Battle of the Boyne

1690 The English philosopher John Locke (1632–1704) produces his *Two Treatises of Civil Government*, which supports natural law above the authority of a ruling body

1697 Tsar Peter the Great of Russia (1672–1725) visits Europe unannounced; the trip encourages him to Westernize his own country

1700

1703 Tsar Peter the Great founds St. Petersburg, the new capital city, on the northwest coast of Russia

1704 Newton rejects the wave theory of light, presenting his work on light and color in his book *Opticks*

1705 English astronomer Edmund Halley (1656–1742) applies Newton's ideas to comets, correctly predicting the return of one—later to be called "Halley's comet"—on its orbit around the Sun

1710 The Royal Chapel at Versailles, designed by Jules Hardouin Mansart (1645–1708), chief architect to French king Louis XIV (1638–1715), is completed

1710

1712 In the wake of strong pressure from Newton, Flamsteed's star catalogs—charting the position of nearly 3,000 stars—are published without his permission. Newton uses the data for his lunar theory

1715 The first performance of the *Water Music* by George Friederic Handel (1685–1759) takes place on the River Thames in London, England

1715 Louis XV (1710–1774), known as "Louis the Well-Beloved," is crowned king in France

1719 *Robinson Crusoe* by Daniel Defoe (1660–1731) is the first known English novel. Based on a true story, it describes the life of a man wrecked on a remote island

1720

1720 After Flamsteed's death, Halley becomes astronomer-royal at the Greenwich Observatory

1724 Tsar Peter the Great (1672–1725) founds the Academy of Sciences in Saint Petersburg, Russia

1726 English writer and clergyman Jonathan Swift (1667–1745) completes his satire on politics, *Gulliver's Travels*. Among the characters that Lemuel Gulliver meets are the inhabitants of Lilliput, who are only six inches high

1730

1732 American statesman and scientist Benjamin Franklin (1706–1790) begins publication of *Poor Richard's Almanac*, which gains a vast circulation in the American colonies

After 1730

1748 Swiss mathematician Leonhard Euler (1707–1783) publishes *Introduction to Infinitesimal Analysis;* later produces textbooks on differential and integral calculus

1905, 1915 German-American physicist Albert Einstein (1879–1955) presents his special and general theories of relativity (1905 and 1915), in which he revises aspects of Newton's laws on motion and gravity

PERIODIC TABLE OF ELEMENTS

The periodic table organizes all the chemical elements into a simple chart according to the physical and chemical properties of their atoms. The elements are arranged by atomic number from 1 to 116. The atomic number is based on the number of protons in the nucleus of the atom. The atomic mass is the combined mass of protons and neutrons in the nucleus. Each element has a chemical symbol that is an abbreviation of its name. In some cases, such as potassium,

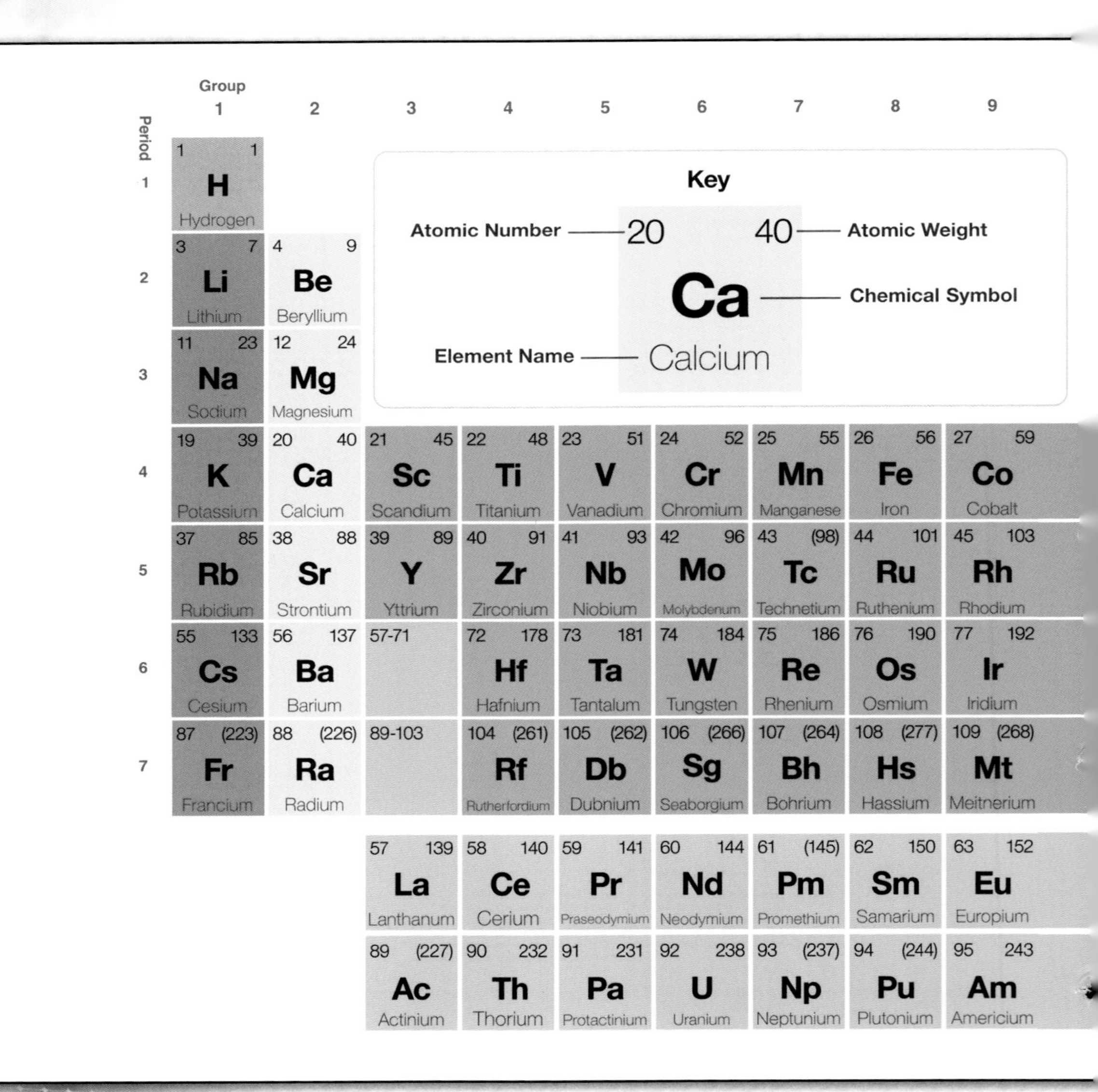

the symbol is an abbreviation of its Latin name ("K" stands for kalium). The name by which the element is commonly known is given in full underneath the symbol. The last item in the element box is the atomic mass. This is the average mass of an atom of the element.

Scientists have arranged the elements into vertical columns called groups and horizontal rows called periods. Elements in any one group all have the same number of electrons in their outer shell and have similar chemical properties. Periods represent the increasing number of electrons it takes to fill the inner and outer shells and become stable. When all the spaces have been filled (Group 18 atoms have all their shells filled), the next period begins. Further explanation of the periodic table is given in Volume 5.

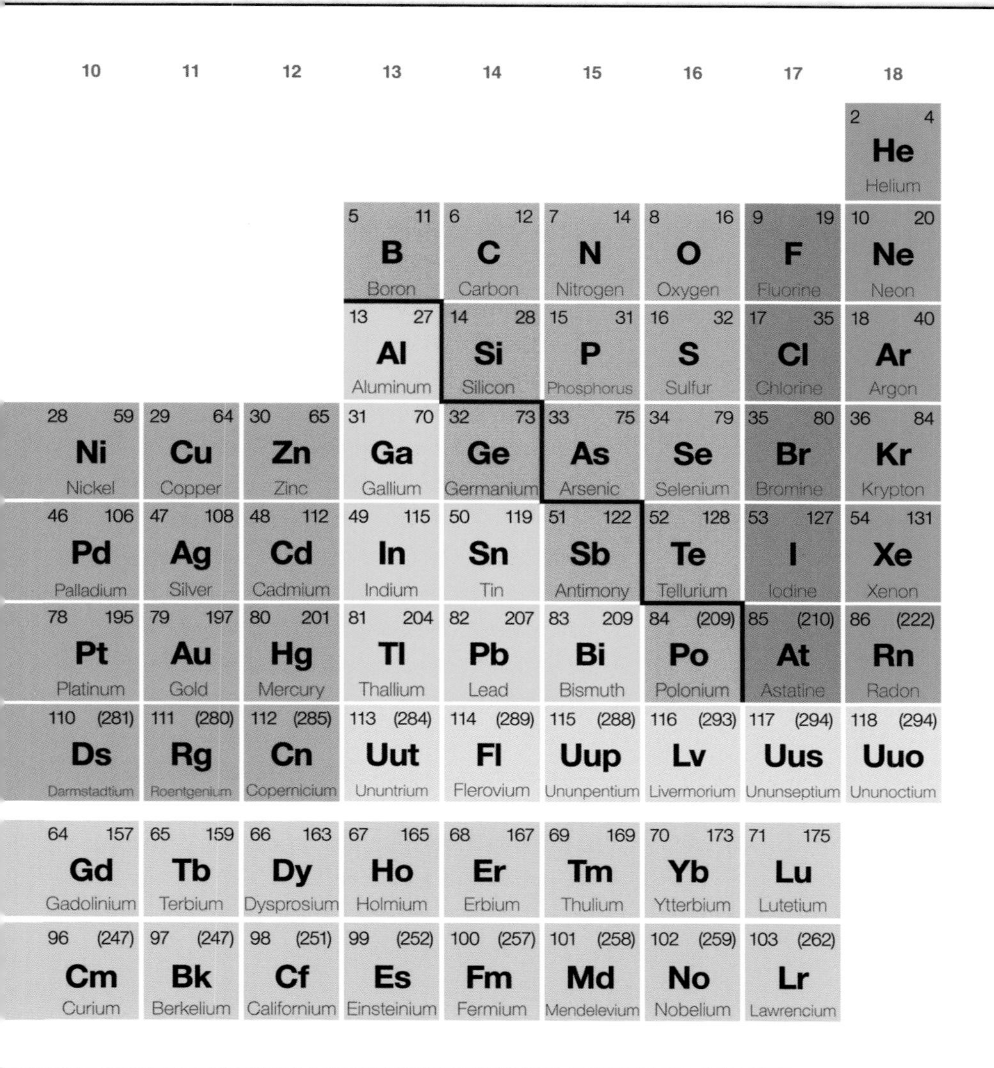

10	11	12	13	14	15	16	17	18
								2 4 He Helium
			5 11 B Boron	6 12 C Carbon	7 14 N Nitrogen	8 16 O Oxygen	9 19 F Fluorine	10 20 Ne Neon
			13 27 Al Aluminum	14 28 Si Silicon	15 31 P Phosphorus	16 32 S Sulfur	17 35 Cl Chlorine	18 40 Ar Argon
28 59 Ni Nickel	29 64 Cu Copper	30 65 Zn Zinc	31 70 Ga Gallium	32 73 Ge Germanium	33 75 As Arsenic	34 79 Se Selenium	35 80 Br Bromine	36 84 Kr Krypton
46 106 Pd Palladium	47 108 Ag Silver	48 112 Cd Cadmium	49 115 In Indium	50 119 Sn Tin	51 122 Sb Antimony	52 128 Te Tellurium	53 127 I Iodine	54 131 Xe Xenon
78 195 Pt Platinum	79 197 Au Gold	80 201 Hg Mercury	81 204 Tl Thallium	82 207 Pb Lead	83 209 Bi Bismuth	84 (209) Po Polonium	85 (210) At Astatine	86 (222) Rn Radon
110 (281) Ds Darmstadtium	111 (280) Rg Roentgenium	112 (285) Cn Copernicium	113 (284) Uut Ununtrium	114 (289) Fl Flerovium	115 (288) Uup Ununpentium	116 (293) Lv Livermorium	117 (294) Uus Ununseptium	118 (294) Uuo Ununoctium
64 157 Gd Gadolinium	65 159 Tb Terbium	66 163 Dy Dysprosium	67 165 Ho Holmium	68 167 Er Erbium	69 169 Tm Thulium	70 173 Yb Ytterbium	71 175 Lu Lutetium	
96 (247) Cm Curium	97 (247) Bk Berkelium	98 (251) Cf Californium	99 (252) Es Einsteinium	100 (257) Fm Fermium	101 (258) Md Mendelevium	102 (259) No Nobelium	103 (262) Lr Lawrencium	

absolute zero The lowest temperature theoretically possible, –459.67ºF (–273.15ºC).

acid Substance that dissolves in water to form hydrogen ions (H+). Acids are neutralized by alkalis and have a pH below 7.

activated complex A short-lived molecule formed during a reaction that quickly breaks up into either the original reactant molecules or into product molecules.

activation energy The minimum energy needed for reactants to change into products.

active site The place on an enzyme or other catalyst where the reactants attach and the reaction occurs.

adsorption The process of molecules becoming attached to a surface.

alkali Substance that dissolves in water to form hydroxide ions (OH-). Alkalis have a pH greater than 7 and will react with acids to form salts.

allotrope A different form of an element in which the atoms are arranged in a different structure.

atom The smallest independent building block of matter. All substances are made of atoms.

atomic mass The number of protons and neutrons in an atom's nucleus.

atomic number The number of protons in a nucleus.

boiling point The temperature at which a liquid turns into a gas.

bond A chemical connection between atoms.

by-product A substance that is produced when another material is made.

calorimeter Apparatus for accurately measuring heat given out or taken in.

catalyst Substance that speeds up a chemical reaction but is left unchanged at the end of the reaction.

chemical equation Symbols and numbers that show how reactants change into products during a chemical reaction.

chemical formula The letters and numbers that represent a chemical compound, such as "H_2O" for water.

chemical reaction The reaction of two or more chemicals (the reactants) to form new chemicals (the products).

chemical symbol The letters that represent a chemical, such as "Cl" for chlorine or "Na" for sodium.

combustion The reaction that causes burning. Combustion is generally a reaction with oxygen in the air.

compound Substance made from more than one element and that has undergone a chemical reaction.

compress To reduce in size or volume by squeezing or exerting pressure.

conductor A substance that carries electricity and heat well.

corrosion The slow wearing away of metals or solids by chemical attack.

covalent bond Bond in which atoms share one or more electrons.

crystal A solid made of regular repeating patterns of atoms.

crystal lattice The arrangement of atoms in a crystalline solid.

density The mass of substance in a unit of volume.

diffuse Spread out evenly, as heat and gases do.

dipole attraction The attractive force between the electrically charged ends of molecules.

elasticity Ability of a material to return to its original shape after being stretched.

electricity A stream of electrons or other charged particles moving through a substance.

electrolyte Liquid containing ions that carries a current between electrodes.

electromagnetic radiation The energy emitted by a source, for example, X-rays, ultraviolet light, visible light, heat, or radio waves.

electromagnetic spectrum The range of energy waves that includes light, heat, and radio waves.

electron A tiny, negatively charged particle that moves around the nucleus of an atom.

electronegativity The power of an atom to attract an electron. Nonmetals, which have only a few spaces in their outer shell, are the most electronegative. Metals, which have several empty spaces, are the least electronegative elements. These metals tend to lose electrons in chemical reactions. Metals of this type are termed electropositive.

element A material that cannot be broken up into simpler ingredients. Elements contain only one type of atom.

endothermic reaction A reaction that absorbs heat energy.

energy The ability to do work.

energy level Electron shells each represent a different energy level. Those closest to the nucleus have the lowest energy.

enthalpy The change in energy during a chemical reaction.

entropy A measure of the amount of disorder or randomness in any system.

enzyme A biological protein that acts as a catalyst.

equilibrium In chemistry, the state of a chemical process in which the products are being formed as fast as they are breaking down, so the amounts of reactant and product are not changing.

evaporation The change of state from a liquid to a gas when the liquid is at a temperature below its boiling point.

exothermic reaction A reaction that releases energy.

fission Process by which a large atom breaks up into two or more smaller fragments.

fusion The process by which two or more small atoms fuse to make a single larger atom.

gas State in which particles are not joined and are free to move in any direction.

heat The transfer of energy between atoms. Adding heat makes atoms move more quickly.

heat capacity The amount of heat required to change the temperature of an object by 1 degree Celsius (1.8°F).

heat of fusion The amount of energy needed to turn a solid into a liquid.

heat of vaporization The amount of energy needed to turn a liquid into a gas.

hydrogen bond A weak dipole attraction that always involves a hydrogen atom.

inhibitor A substance that reduces the effect of a catalyst, causing a catalyzed chemical reaction to slow down.

insulator A substance that does not transfer an electric current or heat.

intermolecular bond Bond that hold molecules together. These bonds are weaker than those between atoms in a molecule.

internal energy The total kinetic energy of all the particles in a system, plus all the chemical energy.

intramolecular bond Strong bond between atoms in a molecule.

ion An atom that has lost or gained one or more electrons.

ionic bond Bond in which one atom gives one or more electrons to another atom.

ionization The formation of ions by adding or removing electrons from atoms.

Kelvin scale Temperature scale that uses degrees kelvin (K) as the unit of temperature, and where zero (0K) is absolute zero (–459.67°F, –273.15°C).

kinetic energy The energy of movement.

kinetic theory The study of heat flow and other processes in terms of the motion of the atoms and molecules involved.

liquid Substance in which particles are loosely bonded and are able to move freely around each other.

matter Anything that can be weighed.

melting point The temperature at which a solid changes into a liquid. When a liquid changes into a solid, this same temperature is called the freezing point.

microstate The state of a substance on the molecular scale—that is, the masses, speeds, and positions of all its molecules.

molecule Two or more joined atoms that have a unique shape and size.

neutron One of the particles that make up the nucleus of an atom. Neutrons do not have any electric charge.

nucleus The central part of an atom. The nucleus contains protons and neutrons except in the case of hydrogen, in which the nucleus contains only one proton.

oxidation The addition of oxygen to a compound.

photon A particle that carries a quantity of energy, such as in the form of light.

potential energy The amount of stored energy that an object possesses because of the way it is positioned or its parts are arranged. A chemical bond has potential energy that is released by breaking the bond.

pressure The force produced by pressing on something.

products The new substance or substances created by a chemical reaction.

proton A positively charged particle found in an atom's nucleus.

radiation The products of radioactivity-alpha and beta particles and gamma rays.

reactants The ingredients necessary for a chemical reaction.

reaction rate The rate at which the concentrations of reactants and products change during the reaction.

relative atomic mass A measure of the mass of an atom compared with the mass of another atom. The values used are the same as those for atomic mass.

relative molecular mass The sum of all the atomic masses of the atoms in a molecule.

reversible reaction A reaction that can go both forward and backward. As well as the reactants forming the products (as in all reactions), the products can react together to re-form the reactants in significant amounts.

shell The orbit of an electron. Each shell can contain a specific maximum number of electrons and no more.

solid State of matter in which particles are held in a rigid framework.

specific heat capacity The amount of heat required to change the temperature of a specified amount of a substance by 1ºC (1.8ºF).

spontaneous reaction A reaction that happens by itself, without needing something else to start it off.

standard conditions Normal room temperature and pressure.

state The form that matter takes—either a solid, a liquid, or a gas.

subatomic particles Particles that are smaller than an atom.

substrate The particular molecule on which an enzyme acts.

temperature A measure of how fast molecules are moving.

thermodynamics The study of how heat and other forms of energy are converted into each other.

valence electrons The electrons in the outer shell of an atom.

van der Waals forces Short-lived forces between atoms and molecules.

volatile Describes a liquid that evaporates easily.

volume The space that a solid, liquid, or gas occupies.

wavelength The distance measured from the peak of one wave to the peak of the next wave.

work The energy used when a force moves an object or changes its shape.

American Association for the Advancement of Science (AAAS)
1200 New York Avenue NW
Washington, DC 20005
Web site: http://www.aaas.org
(202) 326-6400
The mission of the American Association for the Advancement of Science is advancing science around the world by serving as an educator, leader, spokesperson and professional association.

American Association for Clinical Chemistry (AACC)
1850 K Street NW, Suite 625
Washington, DC 20006
(800) 892-1400
Web site: http://www.aacc.org
Founded in 1948 and with 8,000 members, the American Association for Clinical Chemistry is an international scientific/medical society of clinical laboratory professionals, physicians, research scientists, and other individuals involved with clinical chemistry and related disciplines.

American Chemical Society (ACS)
1155 Sixteenth Street NW
Washington, DC 20036
(800) 227-5558
Web site: http://www.acs.org
The American Chemical Society is a congressionally chartered independent membership organization that represents professionals at all degree levels and in all fields of chemistry and sciences that involve chemistry.

American Institute of Chemical Engineers (AIChE)
3 Park Avenue, 19th Fl.
New York, NY 10016-5991
(800) 242-4363
Web site: http://www.aiche.org
The American Institute of Chemical Engineers is the world's leading organization for chemical engineering professionals.

Canadian Society for Chemistry
130 Slater Street, Suite 550
Ottawa, ON K1P 6E2
Canada
(888) 542-2242
Web site: http://www.cheminst.ca
The Canadian Society for Chemistry is a national, not-for-profit, professional association that unites chemistry students and professionals who work in industry, academia and government.

WEB SITES

Due to the changing nature of Internet links, Rosen Publishing has developed an online list of Web sites related to the subject of this book. This site is updated regularly. Please use this link to access the list:

http://www.rosenlinks.com/CORE/Ener

Atkins, P. W. *The Laws of Thermodynamics: A Very Short Introduction.* Oxford, England: Oxford University Press, 2010.

Atkins, P. W. *Reactions: The Private Life of Atoms.* Oxford, England: Oxford University Press, 2011.

Ben-Naim, Arieh. *Entropy and the Second Law: Interpretation and Misss-Interpretationsss.* Hackensack, NJ: World Scientific, 2012.

Brown, T., Burdge, J., Bursten, B., and LeMay, E. *Chemistry: The Central Science.* 10th ed. Englewood Cliffs, NJ: Prentice Hall, 2005.

Cobb, C., and Fetterolf, M. L. *The Joy of Chemistry: The Amazing Science of Familiar Things.* Amherst, NY: Prometheus Books, 2005.

Coopersmith, Jennifer. *Energy, the Subtle Concept: The Discovery of Feynman's Blocks from Leibniz to Einstein.* New York, NY: Oxford University Press, 2010.

Gormley, Beatrice. *Marie Curie: Young Scientist.* New York, NY: Aladdin Paperbacks, 2007.

Greene, B. *Icarus at the Edge of Time.* New York, NY: Alfred A. Knopf, 2008.

Kean, Sam. *The Disappearing Spoon: And Other True Tales of Madness, Love, and the History of the World from the Periodic Table of the Elements.* New York, NY: Back Bay, 2011.

Keeler, James, and Peter Wothers. *Why Chemical Reactions Happen.* Oxford, England: Oxford University Press, 2003.

Lew, Kristi. *Chemical Reactions.* New York, NY: Chelsea House, 2008.

Poynter, M. *Marie Curie: Discoverer of Radium* (Great Minds of Science). Berkeley Heights, NJ: Enslow Publishers, 2007.

Saunders, N. *Fluorine and the Halogens.* Chicago, IL: Heinemann Library, 2005.

Scerri, Eric R. *The Periodic Table: Its Story and Its Significance.* Oxford, England: Oxford University Press, 2007.

VanCleave, Janice Pratt. *Step-by-step Science Experiments in Chemistry.* New York: Rosen Publishing, 2013.

VanCleave, Janice Pratt. *Step-by-step Science Experiments in Energy.* New York: Rosen Publishing, 2013.

PHOTO CREDITS

Cover, p. 3, interior pages (borders) © iStockphoto.com.com/Kativ; pp. 6, 7, 29, 40 (top), 43, 80 iStockphoto/Thinkstock; pp. 8, 76 Apic/Hulton Archive/Getty Images; p. 9 Photo Researchers/Getty Images; p. 10 Science and Society/SuperStock; p. 11 Francoise Sauze/Science Source; p. 12 Nagel Photography/Shutterstock.com; p. 13 (top) Christopher Parypa/Shutterstock.com; p. 13 (bottom) Fritz Goro/Time & Life Pictures/Getty Images; pp. 15, 81 Science & Society Picture Library/Getty Images; p. 16 Flaxphotos/Shutterstock .com; p. 17 Ssnowball/Shutterstock.com; p. 19 Jose Antonio Perez/Shutterstock .com; p. 20 rozbyshaka/Shutterstock.com; p. 21 mihalec/Shutterstock.com; p. 22 crolique/Shutterstock.com; p. 23 Ryan R. Fox/Shutterstock.com; p. 25 (top) Ryan McVay/Photodisc/Thinkstock; p. 25 (bottom) ozgurcankaya/E+/Getty Images; p. 26 grublee/Shutterstock.com; p. 27 Andresr/Shutterstock.com; p. 30 © iStockphoto.com/filo; p. 31 Sinelyov/Shutterstock.com; p. 33 Susanna Price/Dorling Kindersley/Getty Images; p. 34 Tetra Images/Getty Images; p. 36 Mike Powell/Lifesize/Thinkstock; p. 40 (bottom) Darren Hedges/Shutterstock.com; p. 41 Science Source/Photo Researchers/Getty Images; p. 42 Jupiterimages/Comstock/Thinkstock; p. 44 NYPL/Science Source/Photo Researchers/Getty Images; p. 45 (top) Peter Dazeley/Photographer's Choice/Getty Images; p. 45 (bottom) farbled/Shutterstock.com; p. 46 © iStockphoto.com.com/Pr3t3nd3r; p. 48 Albert Barr/Shutterstock.com; p. 50 (top) cosma/Shutterstock.com; p. 50 (bottom) FPG/Archive Photos/Getty Images; pp. 52 Charles D. Winters/Science Source; p. 53 shalunts/Shutterstock.com; p. 57 Andy Sotiriou/The Image Bank/Getty Images; p. 59 Charles D. Winters/Photo Researchers/Getty Images; p. 60 djgis/Shutterstock .com; p. 62 Paul Yates/Shutterstock.com; p. 63 © Henri Martinie/Roger-Viollet/The Image Works; p. 64 Dirk Ercken/Shutterstock.com; p. 65 (left) Epics/Hulton Archive/Getty Images; p. 65 (right) Ingolf Hatz/Cultura/Getty Images; p. 66 (top) Universal Images Group/Getty Images; p. 66 (bottom) Ruslan Rizvanov/Shutterstock.com; p. 67 (top) Denis Tabler/Shutterstock.com; p. 66 (bottom) Malcolm Fielding/Science Source; p. 68 wrangler/Shutterstock.com; p. 69 (top) James King-Holmes/Science Source; p. 69 (bottom) Bloomberg/Getty Images; p. 71 Anatolich/Shutterstock.com; p. 72 Kjersti Joergensen/Shutterstock.com; p. 73 Peter Hince/Photographer's Choice/Getty Images; p. 74 DEA Picture Library/De Agostini/Getty Images; pp. 75 (top), 77 The Bridgeman Art Library/Getty Images; p. 75 (bottom) Photos.com/Thinkstock; pp. 86–87 Kris Everson; top of pp. 6, 23, 36, 52, 64, 74 © iStockphoto.com/aleksandar velasevic; all other photos and illustrations © Brown Bear Books Ltd.